Contractor's Pricing Guide

Framing & Rough Carpentry 2001

Senior Editor
Robert W. Mewis

Contributing Editors
Barbara Balboni
Robert A Bastoni
Howard M. Chandler
John H. Chiang, PE
Paul C. Crosscup
Jennifer L. Curran
Stephen E. Donnelly
J. Robert Lang
Robert C. McNichols
Melville J. Mossman, PE
John J. Moylan
Jeannene D. Murphy
Peter T. Nightingale
Stephen C. Plotner
Michael J. Regan
Stephen J. Sechovicz
Phillip R. Waier, PE

Manager, Engineering Operations
John H. Ferguson, PE

Vice President and General Manager
Roger J. Grant

Group Vice President
Durwood Snead

Vice President, Sales and Marketing
John M. Shea

Production Manager
Michael Kokernak

Production Coordinator
Marion E. Schofield

Technical Support
Thomas J. Dion
Michael H. Donelan
Jonathan Forgit
Mary Lou Geary
Gary L. Hoitt
Paula Reale-Camelio
Robin Richardson
Kathryn S. Rodriguez
Sheryl A. Rose
Elizabeth Ryan
James N. Wills

Book & Cover Design
Norman R. Forgit

Contractor's Pricing Guide

Framing & Rough Carpentry

2001

Published by the R.S. Means Company, Inc.

$36.95 per copy. (In United States).
Price subject to change without prior notice.

R.S. Means Company, Inc.

Construction Publishers & Consultants

Construction Plaza

63 Smiths Lane

Kingston, MA 02364-0800

(781) 585-7880

Printed in the United States of America

ISSN 1074-0473
ISBN 0-87629-601-0
First Printing

Foreword

R.S. Means Co., Inc. is a subsidary of CMD Group (Construction Market Data), a leading provider of construction information, products, and services in North America and globally. CMD's flagship product line includes more than 100 regional editions, national construction project information, sales leads, and over 70 local plan room/libraries in major business centers. CMD also publishes ProFile, a directory of more than 20,000 architectural firms in the U.S. Architect's First Source (AFS) provides Publishers First Source for Products, CSI's SPEC-DATA® and CSI's MANU-SPEC® building product directories for the selection of nationally available building products. R.S. Means provides Construction cost data and training, and consulting services in print, CD-ROM and online. Associated Construction Publications (ACP) provides regional magazines covering the U.S. highways heavy construction, and heavy equipment trader. Manufacturers' Survey Associates (MSA) provides surveys, plans, and specifications. CMD Group is headquartered in Atlanta and has 1,400 employees worldwide. CMD Group is owned by Cahners Business Information (www.cahners.com), a member of the Reed Elsivier plc group (NYSE: RUK and ENL). Cahners is the leading U.S. provider of business information to 16 vertical markets, including building & construction, entertainment, manufacturing, and retail. Cahners' content portfolio encompasses 140 Web sites as well as *Variety, Publishers Weekly, Design News* and 127 other market-leading business-to-business magazines. Cahners developed the leading business-to-business Web portals e-inSITE in the electronics industry, Manufacturing.net in manufacturing as well as Buildingteam.com and HousingZone.com in construction, and maintains an active program of Internet and print launches.

Our Mission

Since 1942, R.S. Means Company, Inc. has been actively engaged in construction cost publishing and consulting throughout North America.

Today, over fifty years after the company began, our primary objective remains the same: to provide you, the construction and facilities professional, with the most current and comprehensive construction cost data possible.

Whether you are a contractor, an owner, an architect, an engineer, a facilities manager, or anyone else who needs a fast and reliable construction cost estimate, you'll find this publication to be a highly useful and necessary tool.

Today, with the constant flow of new construction methods and materials, it's difficult to find the time to look at and evaluate all the different construction cost possibilities. In addition, because labor and material costs keep changing, last year's cost information is not a reliable basis for today's estimate or budget.

That's why so many construction professionals turn to R.S. Means. We keep track of the costs for you, along with a wide range of other key information, from city cost indexes . . . to productivity rates . . . to crew composition . . . to contractor's overhead and profit rates.

R.S. Means performs these functions by collecting data from all facets of the industry, and organizing it in a format that is instantly accessible to you. From the preliminary budget to the detailed unit price estimate, you'll find the data in this book useful for all phases of construction cost determination.

The Staff, the Organization, and Our Services

When you purchase one of R.S. Means' publications, you are in effect hiring the services of a full-time staff of construction and engineering professionals.

Our thoroughly experienced and highly qualified staff works daily at collecting, analyzing, and disseminating comprehensive cost information for your needs. These staff members have years of practical construction experience and engineering training prior to joining the firm. As a result, you can count on them not only for the cost figures, but also for additional background reference information that will help you create a realistic estimate.

The Means organization is always prepared to help you solve construction problems through its five major divisions: Construction and Cost Data Publishing, Electronic Products and Services, Consulting Services, Insurance Division, and Educational Services.

Besides a full array of construction cost estimating books, Means also publishes a number of other reference works for the construction industry. Subjects include construction estimating and project and business management; special topics such as HVAC, roofing, plumbing, and hazardous waste remediation; and a library of facility management references.

In addition, you can access all of our construction cost data through your computer with Means CostWorks 2001 CD-ROM, an electronic tool that offers over 50,000 lines of Means construction cost data, along with assembly and whole building cost data. You can also access Means cost information from our Web site at www.rsmeans.com

What's more, you can increase your knowledge and improve your construction estimating and management performance with a Means Construction Seminar or In-House Training Program. These two-day seminar programs offer unparalleled opportunities for everyone in your organization to get updated on a wide variety of construction-related issues.

Means also is a worldwide provider of construction cost management and analysis services for commercial and government owners and of claims and valuation services for insurers.

In short, R.S. Means can provide you with the tools and expertise for constructing accurate and dependable construction estimates and budgets in a variety of ways.

Robert Snow Means Established a Tradition of Quality That Continues Today

Robert Snow Means spent years building his company, making certain he always delivered a quality product.

Today, at R.S. Means, we do more than talk about the quality of our data and the usefulness of our books. We stand behind all of our data, from historical cost indexes... to construction materials and techniques... to current costs.

If you have any questions about our products or services, please call us toll-free at 1-800-334-3509. Our customer service representatives will be happy to assist you or visit our Web site at www.rsmeans.com

Table of Contents

Table of Contents (continu

Section One
Cost Information

This section provides the cost information necessary to make your construction estimates accurate and efficient. This information is presented in two different formats. The first and most common format is composed of a graphic illustration of the construction system and a list of the building components that are required. Several options for each system are presented. For example, the "Floor Framing Systems" pages show the cost per square foot for 2" x 8" joist framing as well as the 2" x 10" joist framing. In addition, alternate components are shown to allow you to customize the system for your specific project.

These "systems" can also be combined to create an estimate for a more complex project. For example, if you need to replace the exterior wall system below the floor framing of a second story addition, you can consult the "Exterior Wall Framing Systems" pages to calculate those costs per square foot of wall, and then add in the cost of the floor framing. In this manner, the cost for an entire framing project, matched to your specifications for size and proposed design, can be estimated.

The second format lists separately all the building components used in the various systems. This line item format is also used to list the costs for a wide variety of construction materials which may not necessarily be used in any system. For example, the "Insulation" page lists the cost per square foot for various types of building insulation.

The "How to Use" and "Sample Calculation" pages that follow define the terms that appear in the book and will help you to use the information presented.

The cost data provided in this book is both accurate and realistic for putting together an estimate. Anyone using the information provided should, however, put it into proper perspective based upon their own experience and any price fluctuations that may occur at any time in the marketplace.

A good deal of effort has been expended to break each system down into its smallest manageable tasks. Using the systems as they exist, a contractor can confidently prepare an estimate quickly and accurately knowing that the possibility of leaving something out has been eliminated.

Once an estimate has been prepared the contractor has the luxury of analyzing the system and the costs to determine if any adjustments need to be made to increase the potential of winning the job.

How to Use Cost Information

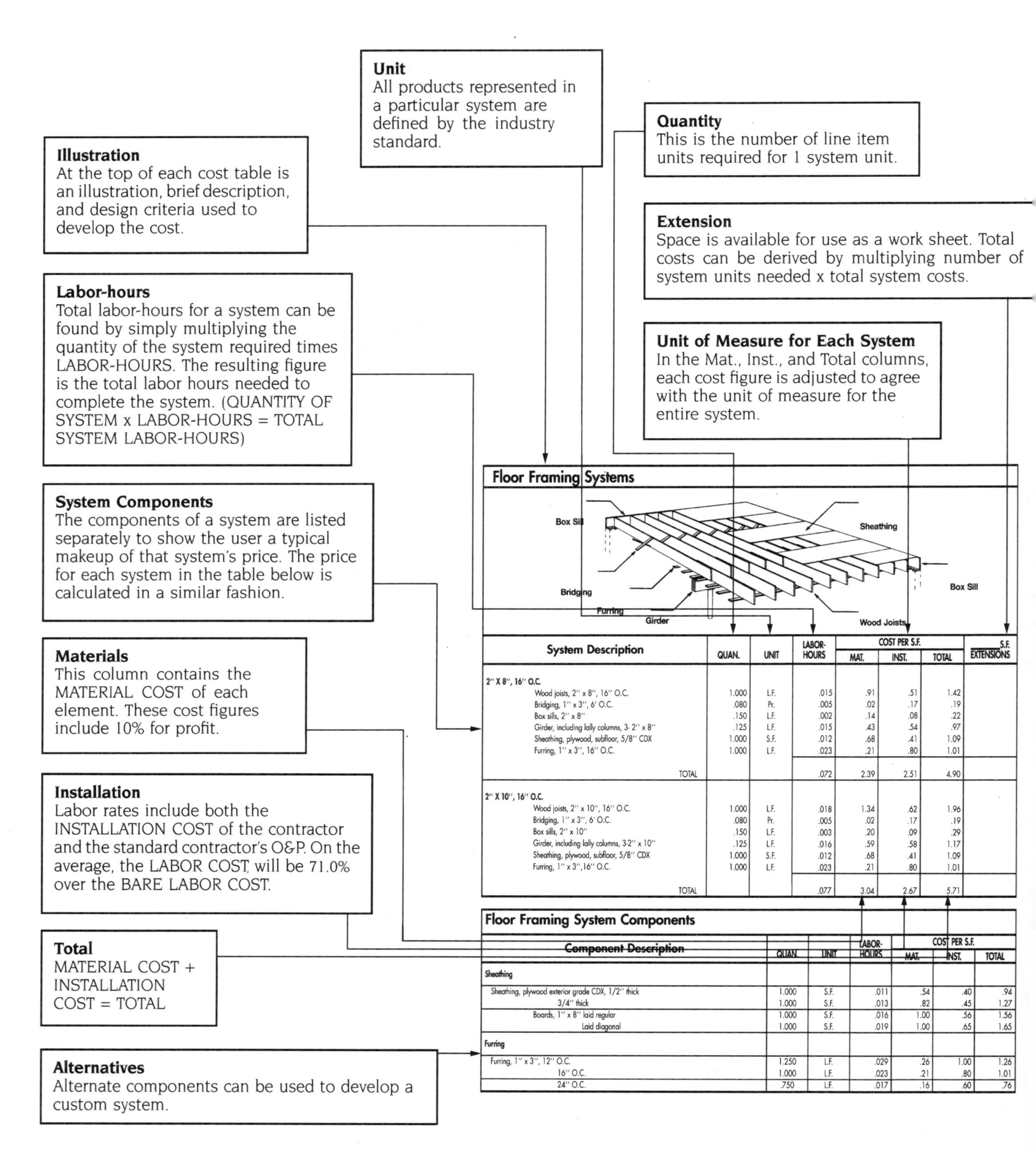

System Description	QUAN.	UNIT	LABOR-HOURS	COST PER S.F. MAT.	COST PER S.F. INST.	COST PER S.F. TOTAL	____S.F. EXTENSIONS
2" X 8", 16" O.C.							
Wood joists, 2" x 8", 16" O.C.	1.000	L.F.	.015	.91	.51	1.42	
Bridging, 1" x 3", 6' O.C.	.080	Pr.	.005	.02	.17	.19	
Box sills, 2" x 8"	.150	L.F.	.002	.14	.08	.22	
Girder, including lally columns, 3- 2" x 8"	.125	L.F.	.015	.43	.54	.97	
Sheathing, plywood, subfloor, 5/8" CDX	1.000	S.F.	.012	.68	.41	1.09	
Furring, 1" x 3", 16" O.C.	1.000	L.F.	.023	.21	.80	1.01	
TOTAL			.072	2.39	2.51	4.90	
2" X 10", 16" O.C.							
Wood joists, 2" x 10", 16" O.C.	1.000	L.F.	.018	1.34	.62	1.96	
Bridging, 1" x 3", 6' O.C.	.080	Pr.	.005	.02	.17	.19	
Box sills, 2" x 10"	.150	L.F.	.003	.20	.09	.29	
Girder, including lally columns, 3-2" x 10"	.125	L.F.	.016	.59	.58	1.17	
Sheathing, plywood, subfloor, 5/8" CDX	1.000	S.F.	.012	.68	.41	1.09	
Furring, 1" x 3",16" O.C.	1.000	L.F.	.023	.21	.80	1.01	
TOTAL			.077	3.04	2.67	5.71	

Floor Framing System Components

Component Description	QUAN.	UNIT	LABOR-HOURS	COST PER S.F. MAT.	COST PER S.F. INST.	COST PER S.F. TOTAL
Sheathing						
Sheathing, plywood exterior grade CDX, 1/2" thick	1.000	S.F.	.011	.54	.40	.94
3/4" thick	1.000	S.F.	.013	.82	.45	1.27
Boards, 1" x 8" laid regular	1.000	S.F.	.016	1.00	.56	1.56
Laid diagonal	1.000	S.F.	.019	1.00	.65	1.65
Furring						
Furring, 1" x 3", 12" O.C.	1.250	L.F.	.029	.26	1.00	1.26
16" O.C.	1.000	L.F.	.023	.21	.80	1.01
24" O.C.	.750	L.F.	.017	.16	.60	.76

Sample Calculations

A) Determine the cost of the 2nd floor framing system for the building shown in the plan at the right using 2″ x 8″ @ 16″ O.C.

B) Compare costs for systems composed of 2″ x 8″ @ 16″ O.C. versus 2″ x 10″ @ 16″ O.C.

C) How do costs change for a 2″ x 8″ @ 16″ O.C. floor framing system when 3/4″ plywood sheathing is used?

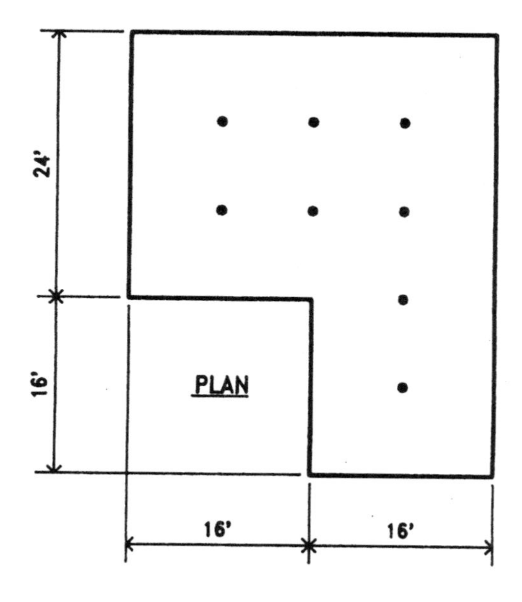

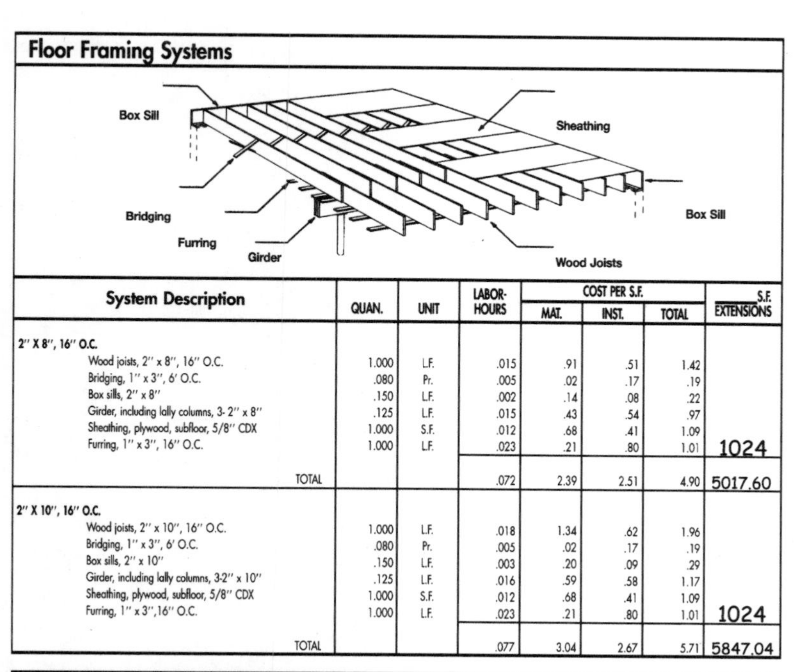

Floor Framing Systems

System Description	QUAN.	UNIT	LABOR-HOURS	COST PER S.F. MAT.	INST.	TOTAL	S.F. EXTENSIONS
2″ X 8″, 16″ O.C.							
Wood joists, 2″ x 8″, 16″ O.C.	1.000	L.F.	.015	.91	.51	1.42	
Bridging, 1″ x 3″, 6′ O.C.	.080	Pr.	.005	.02	.17	.19	
Box sills, 2″ x 8″	.150	L.F.	.002	.14	.08	.22	
Girder, including lally columns, 3-2″ x 8″	.125	L.F.	.015	.43	.54	.97	
Sheathing, plywood, subfloor, 5/8″ CDX	1.000	S.F.	.012	.68	.41	1.09	
Furring, 1″ x 3″, 16″ O.C.	1.000	L.F.	.023	.21	.80	1.01	1024
TOTAL			.072	2.39	2.51	4.90	5017.60
2″ X 10″, 16″ O.C.							
Wood joists, 2″ x 10″, 16″ O.C.	1.000	L.F.	.018	1.34	.62	1.96	
Bridging, 1″ x 3″, 6′ O.C.	.080	Pr.	.005	.02	.17	.19	
Box sills, 2″ x 10″	.150	L.F.	.003	.20	.09	.29	
Girder, including lally columns, 3-2″ x 10″	.125	L.F.	.016	.59	.58	1.17	
Sheathing, plywood, subfloor, 5/8″ CDX	1.000	S.F.	.012	.68	.41	1.09	
Furring, 1″ x 3″,16″ O.C.	1.000	L.F.	.023	.21	.80	1.01	1024
TOTAL			.077	3.04	2.67	5.71	5847.04

Floor Framing System Components

Component Description	QUAN.	UNIT	LABOR-HOURS	COST PER S.F. MAT.	INST.	TOTAL
Sheathing						
Sheathing, plywood exterior grade CDX, 1/2″ thick	1.000	S.F.	.011	.54	.40	.94
3/4″ thick	1.000	S.F.	.013	.82	.45	1.27
Boards, 1″ x 8″ laid regular	1.000	S.F.	.016	1.00	.56	1.56
Laid diagonal	1.000	S.F.	.019	1.00	.65	1.65
Furring						
Furring, 1″ x 3″, 12″ O.C.	1.250	L.F.	.029	.26	1.00	1.26
16″ O.C.	1.000	L.F.	.023	.21	.80	1.01
24″ O.C.	.750	L.F.	.017	.16	.60	.76

Solution:

Determine square footage of floor area to be framed. (24′ x 32′) + (16′ x 16′) = 1,024 S.F.

A) Multiply total square footage by total systems cost per S.F. 1,024 S.F. x $4.90/S.F. = $5,017.60.

B) Costs for 2″ x 10″ @ 16″ O.C. = $5,847.04 (1,024 S.F. x 5.71/S.F.). Difference is $829.44 ($5,847.04 – $5,017.60).

C) Replace cost of 5/8″ sheathing, $1,116.16 (1,024 S.F. x 1.09/S.F.) with cost of 3/4″ sheathing, $1,300.48 (1,024 S.F. x 1.27/S.F.) from systems component page. New systems cost: $5,847.04 – $1,116.16 + $1,300.48 = $6,031.36

Framing Systems

Floor Framing Systems

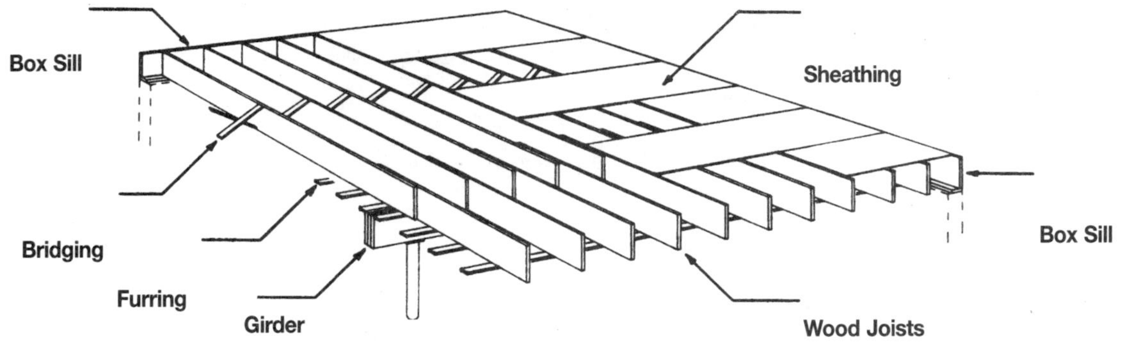

Box Sill
Sheathing
Box Sill
Bridging
Furring
Girder
Wood Joists

System Description	QUAN.	UNIT	LABOR-HOURS	COST PER S.F. MAT.	COST PER S.F. INST.	COST PER S.F. TOTAL	_____S.F. EXTENSIONS
2″ X 6″, 12″ O.C.							
Wood joists, 2″ x 6″, 12″ O.C.	1.250	L.F.	.016	.81	.56	1.37	
Bridging, 1″ x 3″, 6′ OC	.080	Pr.	.005	.02	.17	.19	
Box sills, 2″ x 6″	.150	L.F.	.002	.10	.07	.17	
Girder, including lally columns, 3- 2″ x 8″	.125	L.F.	.015	.43	.54	.97	
Sheathing, plywood, subfloor, 5/8″ CDX	1.000	S.F.	.012	.68	.41	1.09	
Furring, 1″ x 3″, 16″ O.C.	1.000	L.F.	.023	.21	.80	1.01	
TOTAL			.073	2.25	2.55	4.80	
2″ X 6″, 16″ O.C.							
Wood joists, 2″ x 6″, 16″ O.C.	1.000	L.F.	.013	.65	.45	1.10	
Bridging, 1″ x 3″, 6′ OC	.080	Pr.	.005	.02	.17	.19	
Box sills, 2″ x 6″	.150	L.F.	.002	.10	.07	.17	
Girder, including lally columns, 3- 2″ x 8″	.125	L.F.	.015	.43	.54	.97	
Sheathing, plywood, subfloor, 5/8″ CDX	1.000	S.F.	.012	.68	.41	1.09	
Furring, 1″ x 3″, 16″ O.C.	1.000	L.F.	.023	.21	.80	1.01	
TOTAL			.070	2.09	2.44	4.53	
2″ X 8″, 12″ O.C.							
Wood joists, 2″ x 8″, 12″ O.C.	1.250	L.F.	.018	1.14	.64	1.78	
Bridging, 1″ x 3″, 6′ OC	.080	Pr.	.005	.02	.17	.19	
Box sills, 2″ x 8″	.150	L.F.	.002	.14	.08	.22	
Girder, including lally columns, 3- 2″ x 8″	.125	L.F.	.015	.43	.54	.97	
Sheathing, plywood, subfloor, 5/8″ CDX	1.000	S.F.	.012	.68	.41	1.09	
Furring, 1″ x 3″, 16″ O.C.	1.000	L.F.	.023	.21	.80	1.01	
TOTAL			.075	2.62	2.64	5.26	
2″ X 8″, 16″ O.C.							
Wood joists, 2″ x 8″, 16″ O.C.	1.000	L.F.	.015	.91	.51	1.42	
Bridging, 1″ x 3″, 6′ O.C.	.080	Pr.	.005	.02	.17	.19	
Box sills, 2″ x 8″	.150	L.F.	.002	.14	.08	.22	
Girder, including lally columns, 3- 2″ x 8″	.125	L.F.	.015	.43	.54	.97	
Sheathing, plywood, subfloor, 5/8″ CDX	1.000	S.F.	.012	.68	.41	1.09	
Furring, 1″ x 3″, 16″ O.C.	1.000	L.F.	.023	.21	.80	1.01	
TOTAL			.072	2.39	2.51	4.90	

Floor Framing Systems

System Description	QUAN.	UNIT	LABOR-HOURS	COST PER S.F. MAT.	COST PER S.F. INST.	COST PER S.F. TOTAL	____ S.F. EXTENSIONS
2″ X 10″, 12″ O.C.							
Wood joists, 2″ x 10″, 12″ O.C.	1.250	L.F.	.022	1.68	.78	2.46	
Bridging, 1″ x 3″, 6′ OC	.080	Pr.	.005	.02	.17	.19	
Box sills, 2″ x 10″	.150	L.F.	.003	.20	.09	.29	
Girder, including lally columns, 3-2″ x 10″	.125	L.F.	.016	.59	.58	1.17	
Sheathing, plywood, subfloor, 5/8″ CDX	1.000	S.F.	.012	.68	.41	1.09	
Furring, 1″ x 3″,16″ O.C.	1.000	L.F.	.023	.21	.80	1.01	
TOTAL			.081	3.38	2.83	6.21	
2″ X 10″, 16″ O.C.							
Wood joists, 2″ x 10″, 16″ O.C.	1.000	L.F.	.018	1.34	.62	1.96	
Bridging, 1″ x 3″, 6′ O.C.	.080	Pr.	.005	.02	.17	.19	
Box sills, 2″ x 10″	.150	L.F.	.003	.20	.09	.29	
Girder, including lally columns, 3-2″ x 10″	.125	L.F.	.016	.59	.58	1.17	
Sheathing, plywood, subfloor, 5/8″ CDX	1.000	S.F.	.012	.68	.41	1.09	
Furring, 1″ x 3″,16″ O.C.	1.000	L.F.	.023	.21	.80	1.01	
TOTAL			.077	3.04	2.67	5.71	
2″ X 12″, 12″ O.C.							
Wood joists, 2″ x 12″, 12″ O.C.	1.250	L.F.	.023	2.03	.80	2.83	
Bridging, 1″ x 3″, 6′ OC	.080	Pr.	.005	.02	.17	.19	
Box sills, 2″ x 12″	.150	L.F.	.003	.24	.10	.34	
Girder, including lally columns, 3-2″ x 12″	.125	L.F.	.017	.70	.61	1.31	
Sheathing, plywood, subfloor, 5/8″ CDX	1.000	S.F.	.012	.68	.41	1.09	
Furring, 1″ x 3″, 16″ O.C.	1.000	L.F.	.023	.21	.80	1.01	
TOTAL			.083	3.88	2.89	6.77	
2″ X 12″, 16″ O.C.							
Wood joists, 2″ x 12″, 16″ O.C.	1.000	L.F.	.018	1.62	.64	2.26	
Bridging, 1″ x 3″, 6′ O.C.	.080	Pr.	.005	.02	.17	.19	
Box sills, 2″ x 12″	.150	L.F.	.003	.24	.10	.34	
Girder, including lally columns, 3-2″ x 12″	.125	L.F.	.017	.70	.61	1.31	
Sheathing, plywood, subfloor, 5/8″ CDX	1.000	S.F.	.012	.68	.41	1.09	
Furring, 1″ x 3″, 16″ O.C.	1.000	L.F.	.023	.21	.80	1.01	
TOTAL			.078	3.47	2.73	6.20	

Floor costs on this page are given on a cost per square foot basis.

Floor Framing System Components

Component Description	QUAN.	UNIT	LABOR-HOURS	COST PER S.F. MAT.	COST PER S.F. INST.	COST PER S.F. TOTAL
Sheathing						
Sheathing, plywood exterior grade CDX, 1/2″ thick	1.000	S.F.	.011	.54	.40	.94
3/4″ thick	1.000	S.F.	.013	.82	.45	1.27
Boards, 1″ x 8″ laid regular	1.000	S.F.	.016	1.00	.56	1.56
Laid diagonal	1.000	S.F.	.019	1.00	.65	1.65
Furring						
Furring, 1″ x 3″, 12″ O.C.	1.250	L.F.	.029	.26	1.00	1.26
16″ O.C.	1.000	L.F.	.023	.21	.80	1.01
24″ O.C.	.750	L.F.	.017	.16	.60	.76

Floor Framing Systems

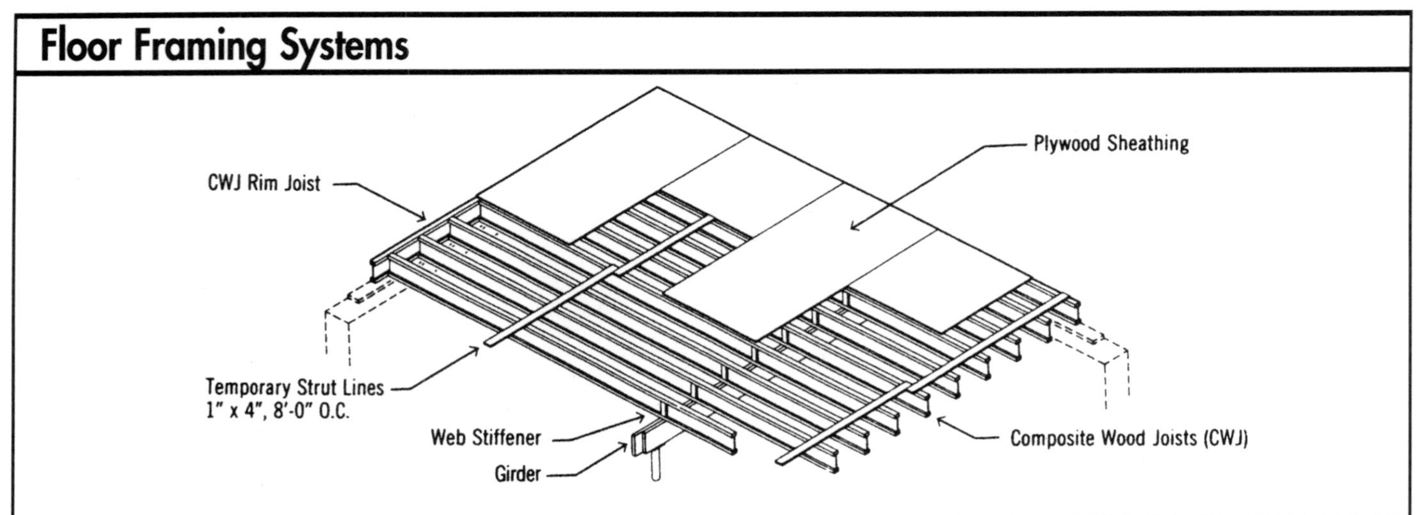

System Description	QUAN.	UNIT	LABOR-HOURS	COST PER S.F. MAT.	COST PER S.F. INST.	COST PER S.F. TOTAL	S.F. EXTENSIONS
9-1/2" COMPOSITE WOOD JOISTS, 16" O.C.							
CWJ, 9-1/2", 16" O.C., 15' span	1.000	L.F.	.018	1.48	.62	2.10	
Temp. strut line, 1" x 4", 8' O.C.	.160	L.F.	.003	.06	.11	.17	
CWJ rim joist, 9-1/2"	.150	L.F.	.003	.22	.09	.31	
Girder, including lally columns, 3- 2" x 8"	.125	L.F.	.015	.43	.54	.97	
Sheathing, plywood, subfloor, 5/8" CDX	1.000	S.F.	.012	.68	.41	1.09	
TOTAL			.051	2.87	1.77	4.64	
9-1/2" COMPOSITE WOOD JOISTS, 24" O.C.							
CWJ, 9-1/2", 24" O.C., 13' span	.750	L.F.	.013	1.11	.47	1.58	
Temp. strut line, 1" x 4", 8' OC	.160	L.F.	.003	.06	.11	.17	
CWJ rim joist, 9-1/2"	.150	L.F.	.003	.22	.09	.31	
Girder, including lally columns, 3- 2" x 10"	.125	L.F.	.015	.43	.54	.97	
Sheathing, plywood, subfloor, 5/8" CDX	1.000	S.F.	.012	.68	.41	1.09	
TOTAL			.046	2.50	1.62	4.12	
11-1/2" COMPOSITE WOOD JOISTS, 16" O.C.							
CWJ, 11-1/2", 16" O.C., 18' span	1.000	L.F.	.018	1.60	.64	2.24	
Temp. strut line, 1" x 4", 8' O.C.	.160	L.F.	.003	.06	.11	.17	
CWJ rim joist, 11-1/2"	.150	L.F.	.003	.24	.10	.34	
Girder, including lally columns, 3-2" x 10"	.125	L.F.	.016	.59	.58	1.17	
Sheathing, plywood, subfloor, 5/8" CDX	1.000	S.F.	.012	.68	.41	1.09	
TOTAL			.052	3.17	1.84	5.01	
11-1/2" COMPOSITE WOOD JOISTS, 24" O.C.							
CWJ, 11-1/2", 24" O.C., 16' span	.750	L.F.	.014	1.20	.48	1.68	
Temp. strut line, 1" x 4", 8' OC	.160	L.F.	.003	.06	.11	.17	
CWJ rim joist, 11-1/2"	.150	L.F.	.003	.24	.10	.34	
Girder, including lally columns, 3- 2" x 10"	.125	L.F.	.016	.59	.58	1.17	
Sheathing, plywood, subfloor, 5/8" CDX	1.000	S.F.	.012	.68	.41	1.09	
TOTAL			.048	2.77	1.68	4.45	

Floor Framing Systems

System Description	QUAN.	UNIT	LABOR-HOURS	COST PER S.F. MAT.	COST PER S.F. INST.	COST PER S.F. TOTAL	S.F. EXTENSIONS
14" COMPOSITE WOOD JOISTS, 16" O.C.							
CWJ, 14", 16" O.C., 22' span	1.000	L.F.	.020	1.80	.68	2.48	
Temp. strut line, 1" x 4", 8' O.C.	.160	L.F.	.003	.06	.11	.17	
CWJ rim joist, 14"	.150	L.F.	.003	.27	.10	.37	
Girder, including lally columns, 3-2" x 12"	.600	L.F.	.017	.70	.61	1.31	
Sheathing, plywood, subfloor, 5/8" CDX	1.000	S.F.	.012	.68	.41	1.09	
TOTAL			.055	3.51	1.91	5.42	
14" COMPOSITE WOOD JOISTS, 24" O.C.							
CWJ, 14", 24" O.C., 19' span	.750	L.F.	.015	1.35	.51	1.86	
Temp. strut line, 1" x 4", 8' OC	.160	L.F.	.003	.06	.11	.17	
CWJ rim joist, 14"	.150	L.F.	.003	.27	.10	.37	
Girder, including lally columns, 3- 2" x 12"	.060	L.F.	.017	.70	.61	1.31	
Sheathing, plywood, subfloor, 5/8" CDX	1.000	S.F.	.012	.68	.41	1.09	
TOTAL			.050	3.06	1.74	4.80	
16" COMPOSITE WOOD JOISTS, 16" O.C.							
CWJ, 16", 16" O.C., 24' span	1.000	L.F.	.018	1.48	.62	2.10	
Temp. strut line, 1" x 4", 8' OC	.160	L.F.	.003	.06	.11	.17	
CWJ rim joist, 16"	.150	L.F.	.003	.35	.11	.46	
Girder, including lally columns, 3- 2" x 12"	.060	L.F.	.017	.70	.61	1.31	
Sheathing, plywood, subfloor, 5/8" CDX	1.000	S.F.	.012	.68	.41	1.09	
TOTAL			.053	3.27	1.86	5.13	
16" COMPOSITE WOOD JOISTS, 24" O.C.							
CWJ, 16", 24" O.C., 22' span	.750	L.F.	.013	1.11	.47	1.58	
Temp. strut line, 1" x 4", 8' OC	.160	L.F.	.003	.06	.11	.17	
CWJ rim joist, 16"	.150	L.F.	.003	.35	.11	.46	
Girder, including lally columns, 3- 2" x 12"	.060	L.F.	.017	.70	.61	1.31	
Sheathing, plywood, subfloor, 5/8" CDX	1.000	S.F.	.012	.68	.41	1.09	
TOTAL			.048	2.90	1.71	4.61	

Floor costs on this page are given on a cost per square foot basis.

Floor Framing System Components

Component Description	QUAN.	UNIT	LABOR-HOURS	COST PER S.F. MAT.	COST PER S.F. INST.	COST PER S.F. TOTAL
Sheathing						
Sheathing, plywood exterior grade CDX, 1/2" thick	1.000	S.F.	.011	.54	.40	.94
5/8" thick	1.000	S.F.	.012	.68	.41	1.09
3/4" thick	1.000	S.F.	.013	.82	.45	1.27
Boards, 1" x 8" laid regular	1.000	S.F.	.016	1.00	.56	1.56
Laid diagonal	1.000	S.F.	.019	1.00	.65	1.65
1" x 10" laid regular	1.000	S.F.	.015	1.11	.51	1.62
Laid diagonal	1.000	S.F.	.018	1.12	.62	1.74
Furring, 1" x 3", 12" O.C.	1.250	L.F.	.029	.26	1.00	1.26
16" O.C.	1.000	L.F.	.023	.21	.80	1.01
24" O.C.	.750	L.F.	.017	.16	.60	.76

Floor Framing Systems

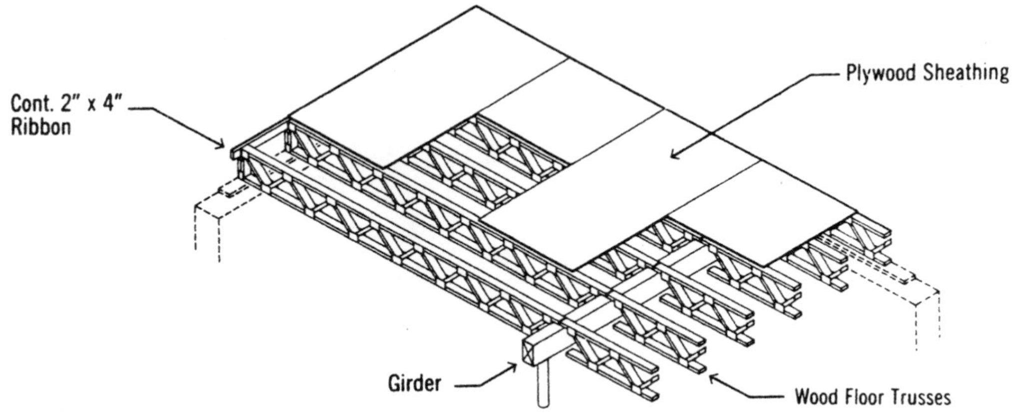

Cont. 2" x 4" Ribbon

Plywood Sheathing

Girder

Wood Floor Trusses

System Description	QUAN.	UNIT	LABOR-HOURS	COST PER S.F. MAT.	COST PER S.F. INST.	COST PER S.F. TOTAL	____S.F. EXTENSIONS
12" OPEN WEB JOISTS, 16" O.C.							
OWJ 12", 16" O.C., 21' span	1.000	L.F.	.018	1.65	.64	2.29	
Continuous ribbing, 2" x 4"	.150	L.F.	.002	.07	.07	.14	
Girder, including lally columns, 3- 2" x 8"	.125	L.F.	.015	.43	.54	.97	
Sheathing, plywood, subfloor, 5/8" CDX	1.000	S.F.	.012	.68	.41	1.09	
Furring, 1" x 3", 16" O.C.	1.000	L.F.	.023	.21	.80	1.01	
TOTAL			.070	3.04	2.46	5.50	
12" OPEN WEB WOOD JOISTS, 24" O.C.							
OWJ, 12", 24" O.C., 17' span	.750	L.F.	.014	1.24	.48	1.72	
Continuous ribbing, 2" x 4"	.150	L.F.	.002	.07	.07	.14	
Girder, including lally columns, 3-2" x 10"	.125	L.F.	.015	.43	.54	.97	
Sheathing, plywood, subfloor, 5/8" CDX	1.000	S.F.	.012	.68	.41	1.09	
Furring, 1" x 3", 16" O.C.	1.000	L.F.	.023	.21	.80	1.01	
TOTAL			.066	2.63	2.30	4.93	
14" OPEN WEB WOOD JOISTS, 16" O.C.							
OWJ 14", 16" O.C., 22' span	1.000	L.F.	.020	1.93	.68	2.61	
Continuous ribbing, 2" x 4"	.150	L.F.	.002	.07	.07	.14	
Girder, including lally columns, 3-2" x 10"	.125	L.F.	.016	.59	.58	1.17	
Sheathing, plywood, subfloor, 5/8" CDX	1.000	S.F.	.012	.68	.41	1.09	
Furring, 1" x 3",16" O.C.	1.000	L.F.	.023	.21	.80	1.01	
TOTAL			.073	3.48	2.54	6.02	
14" OPEN WEB WOOD JOISTS, 24" O.C.							
OWJ 14", 24" O.C., 18' span	.750	L.F.	.015	1.44	.51	1.95	
Continuous ribbing, 2" x 4"	.150	L.F.	.002	.07	.07	.14	
Girder, including lally columns, 3-2" x 10"	.125	L.F.	.016	.59	.58	1.17	
Sheathing, plywood, subfloor, 5/8" CDX	1.000	S.F.	.012	.68	.41	1.09	
Furring, 1" x 3",16" O.C.	1.000	L.F.	.023	.21	.80	1.01	
TOTAL			.068	2.99	2.37	5.36	

Floor Framing Systems

System Description	QUAN.	UNIT	LABOR-HOURS	COST PER S.F. MAT.	INST.	TOTAL	_____S.F. EXTENSIONS
16″ OPEN WEB WOOD JOISTS, 16″ O.C.							
OWJ 16″, 16″ O.C., 24′ span	1.000	L.F.	.021	2.00	.72	2.72	
Continuous ribbing, 2″ x 4″	.150	L.F.	.002	.07	.07	.14	
Girder, including lally columns, 3-2″ x 12″	.125	L.F.	.017	.70	.61	1.31	
Sheathing, plywood, subfloor, 5/8″ CDX	1.000	S.F.	.012	.68	.41	1.09	
Furring, 1″ x 3″, 16″ O.C.	1.000	L.F.	.023	.21	.80	1.01	
TOTAL			.075	3.66	2.61	6.27	
16″ OPEN WEB WOOD JOISTS, 24″ O.C.							
OWJ 16″, 24″ O.C., 20′ span	.750	L.F.	.015	1.50	.54	2.04	
Continuous ribbing, 2″ x 4″	.150	L.F.	.002	.07	.07	.14	
Girder, including lally columns, 3-2″ x 12″	.125	L.F.	.017	.70	.61	1.31	
Sheathing, plywood, subfloor, 5/8″ CDX	1.000	S.F.	.012	.68	.41	1.09	
Furring, 1″ x 3″, 16″ O.C.	1.000	L.F.	.023	.21	.80	1.01	
TOTAL			.069	3.16	2.43	5.59	
18″ OPEN WEB WOOD JOISTS, 16″ O.C.							
OWJ 18″, 16″ O.C., 24′ span	1.000	L.F.	.022	2.08	.75	2.83	
Continuous ribbing, 2″ x 4″	.150	L.F.	.002	.07	.07	.14	
Girder, including lally columns, 3-2″ x 12″	.125	L.F.	.017	.70	.61	1.31	
Sheathing, plywood, subfloor, 5/8″ CDX	1.000	S.F.	.012	.68	.41	1.09	
Furring, 1″ x 3″, 16″ O.C.	1.000	L.F.	.023	.21	.80	1.01	
TOTAL			.076	3.74	2.64	6.38	
18″ OPEN WEB WOOD JOISTS, 24″ O.C.							
OWJ 18″, 24″ O.C., 20′ span	.750	L.F.	.016	1.56	.56	2.12	
Continuous ribbing, 2″ x 4″	.150	L.F.	.002	.07	.07	.14	
Girder, including lally columns, 3-2″ x 12″	.125	L.F.	.017	.70	.61	1.31	
Sheathing, plywood, subfloor, 5/8″ CDX	1.000	S.F.	.012	.68	.41	1.09	
Furring, 1″ x 3″, 16″ O.C.	1.000	L.F.	.023	.21	.80	1.01	
TOTAL			.070	3.22	2.45	5.67	

Floor costs on this page are given on a cost per square foot basis.

Floor Framing System Components

Component Description	QUAN.	UNIT	LABOR-HOURS	COST PER S.F. MAT.	INST.	TOTAL
Sheathing						
Sheathing, plywood exterior grade CDX, 1/2″ thick	1.000	S.F.	.011	.54	.40	.94
3/4″ thick	1.000	S.F.	.013	.82	.45	1.27
Furring						
Furring, 1″ x 3″, 12″ O.C.	1.250	L.F.	.029	.26	1.00	1.26
24″ O.C.	.750	L.F.	.017	.16	.60	.76

Exterior Wall Framing Systems

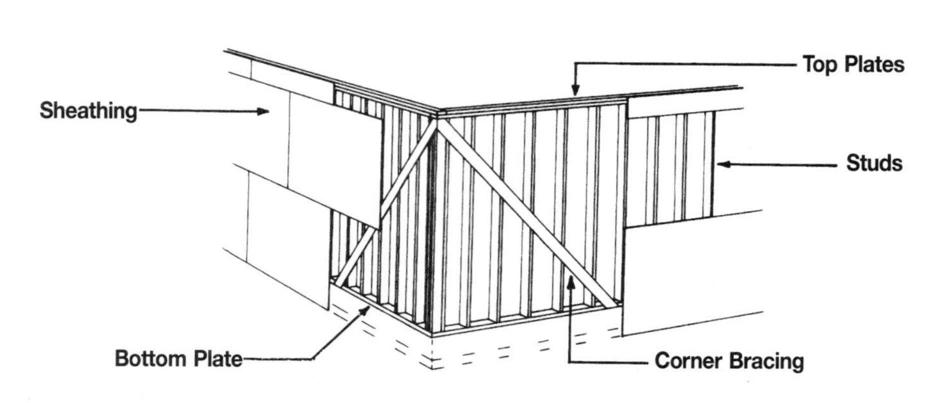

System Description	QUAN.	UNIT	LABOR-HOURS	COST PER S.F. MAT.	COST PER S.F. INST.	COST PER S.F. TOTAL	____ S.F. EXTENSIONS
2" X 4", 12" O.C.							
2" x 4" studs, 12" O.C.	1.250	L.F.	.018	.55	.64	1.19	
Plates, 2" x 4", double top, single bottom	.375	L.F.	.005	.17	.19	.36	
Corner bracing, let-in, 1" x 6"	.063	L.F.	.003	.04	.12	.16	
Sheathing, 1/2" plywood, CDX	1.000	S.F.	.011	.54	.40	.94	
TOTAL			.037	1.30	1.35	2.65	
2" X 4", 16" O.C.							
2" x 4" studs, 16" O.C.	1.000	L.F.	.015	.44	.51	.95	
Plates, 2" x 4", double top, single bottom	.375	L.F.	.005	.17	.19	.36	
Corner bracing, let-in, 1" x 6"	.063	L.F.	.003	.04	.12	.16	
Sheathing, 1/2" plywood, CDX	1.000	S.F.	.011	.54	.40	.94	
TOTAL			.034	1.19	1.22	2.41	
2" X 4", 24" O.C.							
2" x 4" studs, 24" O.C.	.750	L.F.	.011	.33	.38	.71	
Plates, 2" x 4", double top, single bottom	.375	L.F.	.005	.17	.19	.36	
Corner bracing, let-in, 1" x 6"	.063	L.F.	.002	.04	.08	.12	
Sheathing, 1/2" plywood, CDX	1.000	S.F.	.011	.54	.40	.94	
TOTAL			.029	1.08	1.05	2.13	
2" X 6", 12" O.C.							
2" x 6" studs, 12" O.C.	1.250	L.F.	.020	.81	.70	1.51	
Plates, 2" x 6", double top, single bottom	.375	L.F.	.006	.24	.21	.45	
Corner bracing, let-in, 1" x 6"	.063	L.F.	.003	.04	.12	.16	
Sheathing, 1/2" plywood, CDX	1.000	S.F.	.011	.54	.40	.94	
TOTAL			.040	1.63	1.43	3.06	
2" X 6", 16" O.C.							
2" x 6" studs, 16" O.C.	1.000	L.F.	.016	.65	.56	1.21	
Plates, 2" x 6", double top, single bottom	.375	L.F.	.006	.24	.21	.45	
Corner bracing, let-in, 1" x 6"	.063	L.F.	.003	.04	.12	.16	
Sheathing, 1/2" plywood, CDX	1.000	S.F.	.011	.54	.40	.94	
TOTAL			.036	1.47	1.29	2.76	

Exterior Wall Framing Systems

System Description	QUAN.	UNIT	LABOR-HOURS	COST PER S.F. MAT.	COST PER S.F. INST.	COST PER S.F. TOTAL	____S.F. EXTENSIONS
2" X 6", 24" O.C.							
2" x 6" studs, 24" O.C.	.750	L.F.	.012	.49	.42	.91	
Plates, 2" x 6", double top, single bottom	.375	L.F.	.006	.24	.21	.45	
Corner bracing, let-in, 1" x 6"	.063	L.F.	.002	.04	.08	.12	
Sheathing, 1/2" plywood, CDX	1.000	S.F.	.011	.54	.40	.94	
TOTAL			.031	1.31	1.11	2.42	
2" X 8", 16" O.C.							
2" x 8" studs, 16" O.C.	1.000	L.F.	.020	1.24	.70	1.94	
Plates, 2" x 8", double top, single bottom	.375	L.F.	.008	.47	.26	.73	
Corner bracing, let-in, 1" x 6"	.063	L.F.	.002	.04	.08	.12	
Sheathing, 1/2" plywood, CDX	1.000	S.F.	.011	.54	.40	.94	
TOTAL			.041	2.29	1.44	3.73	
2" X 8", 24" O.C.							
2" x 8" studs, 24" O.C.	.750	L.F.	.015	.93	.53	1.46	
Plates, 2" x 8", double top, single bottom	.375	L.F.	.008	.47	.26	.73	
Corner bracing, let-in, 1" x 6"	.063	L.F.	.002	.04	.08	.12	
Sheathing, 1/2" plywood, CDX	1.000	S.F.	.011	.54	.40	.94	
TOTAL			.036	1.98	1.27	3.25	

The wall costs on this page are given in cost per square foot of wall.
For window and door openings see below.

Exterior Wall Framing Components

Component Description	QUAN.	UNIT	LABOR-HOURS	COST PER S.F. MAT.	COST PER S.F. INST.	COST PER S.F. TOTAL
Sheathing						
Sheathing, plywood CDX, 3/8" thick	1.000	S.F.	.010	.41	.37	.78
5/8" thick	1.000	S.F.	.012	.68	.43	1.11
3/4" thick	1.000	S.F.	.013	.82	.46	1.28
Wood fiber, regular, no vapor barrier, 1/2" thick	1.000	S.F.	.013	.58	.46	1.04
5/8" thick	1.000	S.F.	.013	.78	.46	1.24
Asphalt impregnated 25/32" thick	1.000	S.F.	.013	.39	.46	.85
1/2" thick	1.000	S.F.	.013	.32	.46	.78

Window & Door Openings

Component Description	QUAN.	UNIT	LABOR-HOURS	COST EACH MAT.	COST EACH INST.	COST EACH TOTAL
The following costs are to be added to the total costs of the wall for each opening. Do not subtract the area of the openings.						
Headers						
Headers, 2" x 6" double, 2' long	4.000	L.F.	.178	2.60	6.20	8.80
4' long	8.000	L.F.	.356	5.20	12.40	17.60
2" x 8" double, 4' long	8.000	L.F.	.376	7.30	13.10	20.40
6' long	12.000	L.F.	.565	10.90	19.70	30.60
2" x 10" double, 4' long	8.000	L.F.	.400	10.70	13.90	24.60
6' long	12.000	L.F.	.600	16.10	21.00	37.10

Gable End Roof Framing Systems

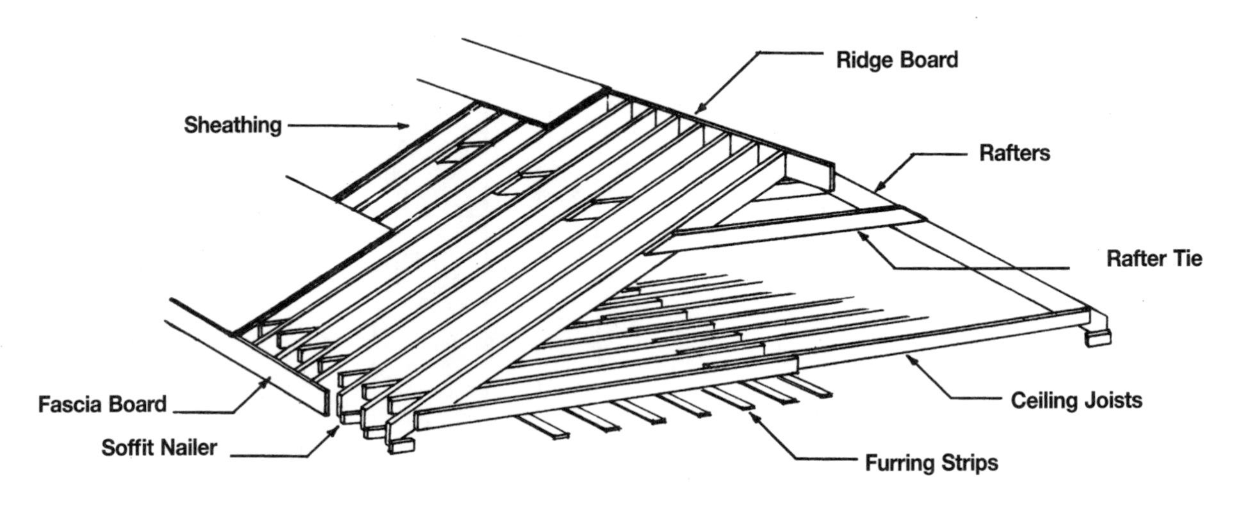

System Description	QUAN.	UNIT	LABOR-HOURS	COST PER S.F. MAT.	COST PER S.F. INST.	COST PER S.F. TOTAL	____S.F. EXTENSIONS
2″ X 6″ RAFTERS, 16″ O.C., 4/12 PITCH							
Rafters, 2″ x 6″, 16″ O.C., 4/12 pitch	1.170	L.F.	.019	.76	.66	1.42	
Ceiling joists, 2″ x 4″, 16″ O.C.	1.000	L.F.	.013	.44	.45	.89	
Ridge board, 2″ x 6″	.050	L.F.	.002	.03	.06	.09	
Fascia board, 2″ x 6″	.100	L.F.	.005	.07	.19	.26	
Rafter tie, 1″ x 4″, 4′ O.C.	.060	L.F.	.001	.02	.04	.06	
Soffit nailer (outrigger), 2″ x 4″, 24″ O.C.	.170	L.F.	.004	.07	.15	.22	
Sheathing, exterior, plywood, CDX, 1/2″ thick	1.170	S.F.	.013	.63	.47	1.10	
Furring strips, 1″ x 3″, 16″ O.C.	1.000	L.F.	.023	.21	.80	1.01	
TOTAL			.080	2.23	2.82	5.05	
2″ X 8″ RAFTERS, 16″ O.C., 4/12 PITCH							
Rafters, 2″ x 8″, 16″ O.C., 4/12 pitch	1.170	L.F.	.020	1.06	.69	1.75	
Ceiling joists, 2″ x 6″, 16″ O.C.	1.000	L.F.	.013	.65	.45	1.10	
Ridge board, 2″ x 8″	.050	L.F.	.002	.05	.06	.11	
Fascia board, 2″ x 8″	.100	L.F.	.007	.09	.25	.34	
Rafter tie, 1″ x 4″, 4′ O.C.	.060	L.F.	.001	.02	.04	.06	
Soffit nailer (outrigger), 2″ x 4″, 24″ O.C.	.170	L.F.	.004	.07	.15	.22	
Sheathing, exterior, plywood, CDX, 1/2″ thick	1.170	S.F.	.013	.63	.47	1.10	
Furring strips, 1″ x 3″, 16″ O.C.	1.000	L.F.	.023	.21	.80	1.01	
TOTAL			.083	2.78	2.91	5.69	
2″ X 8″ RAFTERS, 24″ O.C., 4/12 PITCH							
Rafters, 2″ x 8″, 24″ O.C., 4/12 pitch	.940	L.F.	.016	.86	.55	1.41	
Ceiling joists, 2″ x 6″, 24″ O.C.	.750	L.F.	.010	.49	.34	.83	
Ridge board, 2″ x 8″	.050	L.F.	.002	.03	.06	.09	
Fascia board, 2″ x 8″	.100	L.F.	.005	.07	.19	.26	
Rafter tie, 1″ x 4″, 4′ OC	.060	L.F.	.001	.02	.04	.06	
Soffit nailer (outrigger), 2″ x 4″, 24″ O.C.	.170	L.F.	.004	.07	.15	.22	
Sheathing, exterior, plywood, CDX, 1/2″ thick	1.170	S.F.	.013	.63	.47	1.10	
Furring strips, 1″ x 3″, 16″ O.C.	1.000	L.F.	.023	.21	.80	1.01	
TOTAL			.074	2.38	2.60	4.98	

Gable End Roof Framing Systems

System Description	QUAN.	UNIT	LABOR-HOURS	COST PER S.F. MAT.	INST.	TOTAL	S.F. EXTENSIONS
2" X 10" RAFTERS, 16" O.C., 4/12 PITCH							
Rafters, 2" x 10", 16" O.C., 4/12 pitch	1.170	L.F.	.030	1.57	1.03	2.60	
Ceiling joists, 2" x 8", 16" O.C.	1.000	L.F.	.015	.91	.51	1.42	
Ridge board, 2" x 10"	.050	L.F.	.002	.03	.06	.09	
Fascia board, 2" x 10"	.100	L.F.	.005	.07	.19	.26	
Rafter tie, 1" x 4", 4' OC	.060	L.F.	.001	.02	.04	.06	
Soffit nailer (outrigger), 2" x 4", 24" O.C.	.170	L.F.	.004	.07	.15	.22	
Sheathing, exterior, plywood, CDX, 1/2" thick	1.170	S.F.	.013	.63	.47	1.10	
Furring strips, 1" x 3", 16" O.C.	1.000	L.F.	.023	.21	.80	1.01	
TOTAL			.093	3.51	3.25	6.76	
2" X 10" RAFTERS, 24" O.C., 4/12 PITCH							
Rafters, 2" x 10", 24" O.C., 4/12 pitch	.940	L.F.	.024	1.26	.83	2.09	
Ceiling joists, 2" x 8", 24" O.C.	.750	L.F.	.011	.68	.38	1.06	
Ridge board, 2" x 10"	.050	L.F.	.002	.03	.06	.09	
Fascia board, 2" x 10"	.100	L.F.	.005	.07	.19	.26	
Rafter tie, 1" x 4", 4' OC	.060	L.F.	.001	.02	.04	.06	
Soffit nailer (outrigger), 2" x 4", 24" O.C.	.170	L.F.	.004	.07	.15	.22	
Sheathing, exterior, plywood, CDX, 1/2" thick	1.170	S.F.	.013	.63	.47	1.10	
Furring strips, 1" x 3", 16" O.C.	1.000	L.F.	.023	.21	.80	1.01	
TOTAL			.083	2.97	2.92	5.89	

The cost of this system is based on the square foot of plan area.
All quantities have been adjusted accordingly.

Gable End Roof Framing Components

Component Description	QUAN.	UNIT	LABOR-HOURS	COST PER S.F. MAT.	INST.	TOTAL
Rafters						
Rafters, #2 or better, 16" O.C., 2" x 6", 4/12 pitch	1.170	L.F.	.019	.76	.66	1.42
8/12 pitch	1.330	L.F.	.027	.86	.93	1.79
2" x 8", 4/12 pitch	1.170	L.F.	.020	1.06	.69	1.75
8/12 pitch	1.330	L.F.	.028	1.21	.98	2.19
24" O.C., 2" x 6", 4/12 pitch	.940	L.F.	.015	.61	.53	1.14
8/12 pitch	1.060	L.F.	.021	.69	.74	1.43
2" x 8", 4/12 pitch	.940	L.F.	.016	.86	.55	1.41
8/12 pitch	1.060	L.F.	.023	.96	.78	1.74
Ceiling joists						
Ceiling joist, #2 or better, 2" x 4", 16" O.C.	1.000	L.F.	.013	.44	.45	.89
24" O.C.	.750	L.F.	.010	.33	.34	.67
2" x 10", 16" O.C.	1.000	L.F.	.018	1.34	.62	1.96
24" O.C.	.750	L.F.	.013	1.01	.47	1.48
Ridge board						
Ridge board, #2 or better, 1" x 6"	.050	L.F.	.001	.05	.05	.10
1" x 8"	.050	L.F.	.001	.07	.05	.12
1" x 10"	.050	L.F.	.002	.08	.06	.14
2" x 6"	.050	L.F.	.002	.03	.06	.09
Fascia board						
Fascia board, #2 or better, 1" x 6"	.100	L.F.	.004	.05	.14	.19
1" x 8"	.100	L.F.	.005	.06	.16	.22

Truss Roof Framing Systems

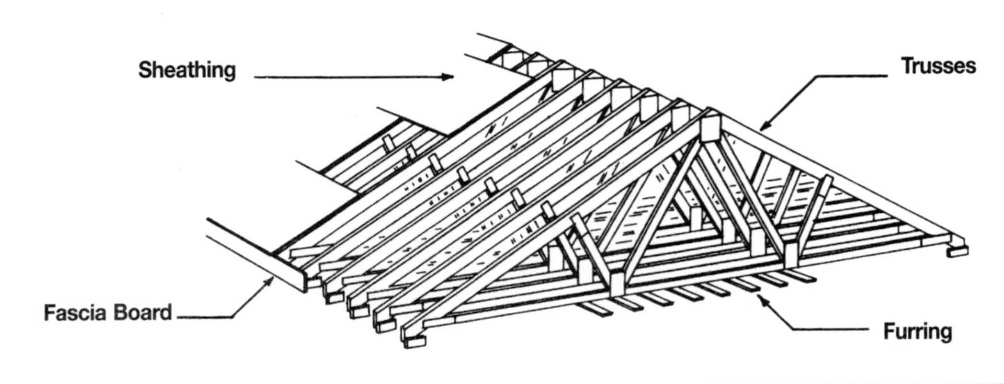

System Description	QUAN.	UNIT	LABOR-HOURS	COST PER S.F. MAT.	COST PER S.F. INST.	COST PER S.F. TOTAL	S.F. EXTENSIONS
TRUSS, 16" O.C., 4/12 PITCH, 1' OVERHANG, 26' SPAN							
Truss, 40# loading, 16" O.C., 4/12 pitch, 26' span	.030	Ea.	.021	2.04	.96	3.00	
Fascia board, 2" x 6"	.100	L.F.	.005	.07	.19	.26	
Sheathing, exterior, plywood, CDX, 1/2" thick	1.170	S.F.	.013	.63	.47	1.10	
Furring, 1" x 3", 16" O.C.	1.000	L.F.	.023	.21	.80	1.01	
TOTAL			.062	2.95	2.42	5.37	
TRUSS, 16" O.C., 8/12 PITCH, 1' OVERHANG, 26' SPAN							
Truss, 40# loading, 16" O.C., 8/12 pitch, 26' span	.030	Ea.	.023	2.42	1.05	3.47	
Fascia board, 2" x 6"	.100	L.F.	.005	.07	.19	.26	
Sheathing, exterior, plywood, CDX, 1/2" thick	1.330	S.F.	.015	.72	.53	1.25	
Furring, 1" x 3", 16" O.C.	1.000	L.F.	.023	.21	.80	1.01	
TOTAL			.066	3.42	2.57	5.99	
TRUSS, 24" O.C., 4/12 PITCH, 1' OVERHANG, 26' SPAN							
Truss, 40# loading, 24" O.C., 4/12 pitch, 26' span	.020	Ea.	.014	1.36	.64	2.00	
Fascia board, 2" x 6"	.100	L.F.	.005	.07	.19	.26	
Sheathing, exterior, plywood, CDX, 1/2" thick	1.170	S.F.	.013	.63	.47	1.10	
Furring, 1" x 3", 16" O.C.	1.000	L.F.	.023	.21	.80	1.01	
TOTAL			.055	2.27	2.10	4.37	
TRUSS, 24" O.C., 8/12 PITCH, 1' OVERHANG, 26' SPAN							
Truss, 40# loading, 24" O.C., 8/12 pitch, 26' span	.020	Ea.	.015	1.61	.70	2.31	
Fascia board, 2" x 6"	.100	L.F.	.005	.07	.19	.26	
Sheathing, exterior, plywood, CDX, 1/2" thick	1.330	S.F.	.015	.72	.53	1.25	
Furring, 1" x 3", 16" O.C.	1.000	L.F.	.023	.21	.80	1.01	
TOTAL			.058	2.61	2.22	4.83	
TRUSS, 16" O.C., 4/12 PITCH, 1' OVERHANG, 32' SPAN							
Truss, 40# loading, 16" O.C., 4/12 pitch, 32' span	.024	Ea.	.019	1.99	.87	2.86	
Fascia board, 2" x 6"	.100	L.F.	.005	.07	.19	.26	
Sheathing, exterior, plywood, CDX, 1/2" thick	1.170	S.F.	.015	.72	.53	1.25	
Furring, 1" x 3", 16" O.C.	1.000	L.F.	.023	.21	.80	1.01	
TOTAL			.062	2.99	2.39	5.38	

Truss Roof Framing Systems

System Description	QUAN.	UNIT	LABOR-HOURS	COST PER S.F. MAT.	COST PER S.F. INST.	COST PER S.F. TOTAL	____S.F. EXTENSIONS
TRUSS, 16″ O.C., 8/12 PITCH, 1′ OVERHANG, 32′ SPAN							
Truss, 40# loading, 16″ O.C., 8/12 pitch, 32′ span	.024	Ea.	.021	2.45	.96	3.41	
Fascia board, 2″ x 6″	.100	L.F.	.005	.07	.19	.26	
Sheathing, exterior, plywood, CDX, 1/2″ thick	1.330	S.F.	.015	.72	.53	1.25	
Furring, 1″ x 3″, 16″ O.C.	1.000	L.F.	.023	.21	.80	1.01	
TOTAL			.064	3.45	2.48	5.93	
TRUSS, 24″ O.C., 4/12 PITCH, 1′ OVERHANG, 32′ SPAN							
Truss, 40# loading, 24″ O.C., 4/12 pitch, 32′ span	.016	Ea.	.013	1.33	.58	1.91	
Fascia board, 2″ x 6″	.100	L.F.	.005	.07	.19	.26	
Sheathing, exterior, plywood, CDX, 1/2″ thick	1.170	S.F.	.015	.72	.53	1.25	
Furring, 1″ x 3″, 16″ O.C.	1.000	L.F.	.023	.21	.80	1.01	
TOTAL			.056	2.33	2.10	4.43	
TRUSS, 24″ O.C., 8/12 PITCH, 1′ OVERHANG, 32′ SPAN							
Truss, 40# loading, 24″ O.C., 8/12 pitch, 32′ span	.016	Ea.	.014	1.63	.64	2.27	
Fascia board, 2″ x 6″	.100	L.F.	.005	.07	.19	.26	
Sheathing, exterior, plywood, CDX, 1/2″ thick	1.330	S.F.	.015	.72	.53	1.25	
Furring, 1″ x 3″, 16″ O.C.	1.000	L.F.	.023	.21	.80	1.01	
TOTAL			.057	2.63	2.16	4.79	

The cost of this system is based on the square foot of plan area.
A one foot overhang is included.

Truss Roof Framing System Components

Component Description	QUAN.	UNIT	LABOR-HOURS	COST PER S.F. MAT.	COST PER S.F. INST.	COST PER S.F. TOTAL
Trusses						
Truss, 40# loading, including 1′ overhang, 4/12 pitch, 24′ span, 16″ O.C.	.033	Ea.	.022	1.60	1.00	2.60
24″ O.C.	.022	Ea.	.015	1.07	.67	1.74
28′ span, 16″ O.C.	.027	Ea.	.020	1.59	.92	2.51
24″ O.C.	.019	Ea.	.014	1.12	.65	1.77
36′ span, 16″ O.C.	.022	Ea.	.019	2.29	.86	3.15
24″ O.C.	.015	Ea.	.013	1.56	.59	2.15
8/12 pitch, 24′ span, 16″ O.C.	.033	Ea.	.024	2.44	1.08	3.52
24″ O.C.	.022	Ea.	.016	1.63	.73	2.36
28′ span, 16″ O.C.	.027	Ea.	.022	2.34	1.00	3.34
24″ O.C.	.019	Ea.	.016	1.64	.70	2.34
36′ span, 16″ O.C.	.022	Ea.	.021	2.66	.97	3.63
24″ O.C.	.015	Ea.	.015	1.82	.67	2.49

Hip Roof Framing Systems

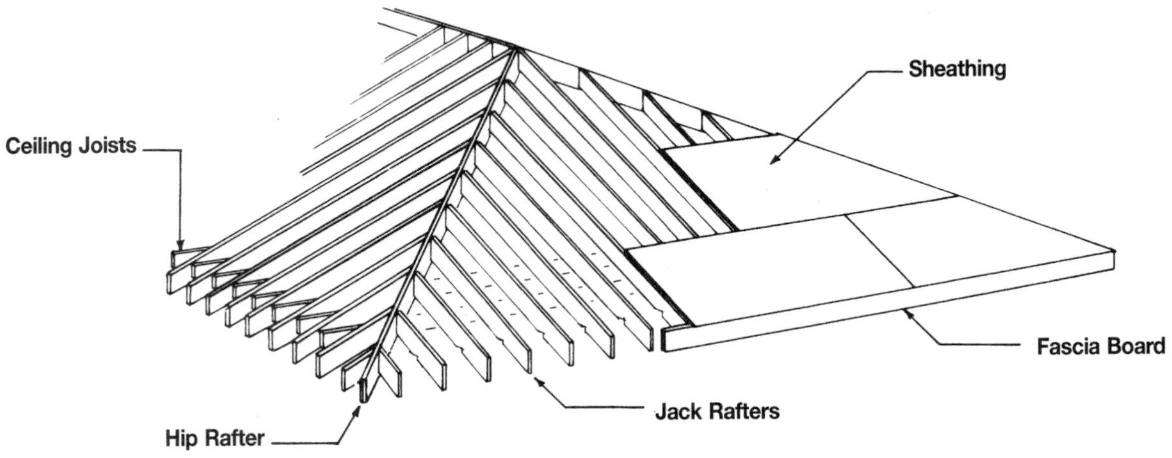

System Description	QUAN.	UNIT	LABOR-HOURS	COST PER S.F. MAT.	COST PER S.F. INST.	COST PER S.F. TOTAL	S.F. EXTENSIONS
2″ X 6″, 16″ O.C., 4/12 PITCH							
Hip rafters, 2″ x 6″, 4/12 pitch	.160	L.F.	.003	.10	.12	.22	
Jack rafters, 2″ x 6″, 16″ O.C., 4/12 pitch	1.430	L.F.	.038	.93	1.33	2.26	
Ceiling joists, 2″ x 4″, 16″ O.C.	1.000	L.F.	.013	.44	.45	.89	
Fascia board, 2″ x 6″	.220	L.F.	.012	.15	.42	.57	
Soffit nailer (outrigger), 2″ x 4″, 24″ O.C.	.220	L.F.	.006	.10	.20	.30	
Sheathing, 1/2″ exterior plywood, CDX	1.570	S.F.	.018	.85	.63	1.48	
Furring strips, 1″ x 3″, 16″ O.C.	1.000	L.F.	.023	.21	.80	1.01	
TOTAL			.113	2.78	3.95	6.73	
2″ X 6″, 24″ O.C., 4/12 PITCH							
Hip rafters, 2″ x 6″, 4/12 pitch	.160	L.F.	.003	.10	.12	.22	
Jack rafters, 2″ x 6″, 24″ O.C., 4/12 pitch	1.150	L.F.	.031	.75	1.07	1.82	
Ceiling joists, 2″ x 4″, 24″ O.C.	.750	L.F.	.010	.33	.34	.67	
Fascia board, 2″ x 6″	.220	L.F.	.016	.20	.54	.74	
Soffit nailer (outrigger), 2″ x 4″, 24″ O.C.	.220	L.F.	.006	.10	.20	.30	
Sheathing, 1/2″ exterior plywood, CDX	1.570	S.F.	.018	.85	.63	1.48	
Furring strips, 1″ x 3″, 16″ O.C.	1.000	L.F.	.023	.21	.80	1.01	
TOTAL			.107	2.54	3.70	6.24	
2″ X 8″, 16″ O.C., 4/12 PITCH							
Hip rafters, 2″ x 8″, 4/12 pitch	.160	L.F.	.004	.15	.12	.27	
Jack rafters, 2″ x 8″, 16″ O.C., 4/12 pitch	1.430	L.F.	.047	1.30	1.63	2.93	
Ceiling joists, 2″ x 6″, 16″ O.C.	1.000	L.F.	.013	.65	.45	1.10	
Fascia board, 2″ x 8″	.220	L.F.	.016	.20	.54	.74	
Soffit nailer (outrigger), 2″ x 4″, 24″ O.C.	.220	L.F.	.006	.10	.20	.30	
Sheathing, 1/2″ exterior plywood, CDX	1.570	S.F.	.018	.85	.63	1.48	
Furring strips, 1″ x 3″, 16″ O.C.	1.000	L.F.	.023	.21	.80	1.01	
TOTAL			.127	3.46	4.37	7.83	

Hip Roof Framing Systems

System Description	QUAN.	UNIT	LABOR-HOURS	COST PER S.F. MAT.	COST PER S.F. INST.	COST PER S.F. TOTAL	S.F. EXTENSIONS
2" X 8", 24" O.C., 4/12 PITCH							
Hip rafters, 2" x 8", 4/12 pitch	.160	L.F.	.003	.10	.12	.22	
Jack rafters, 2" x 8", 24" O.C., 4/12 pitch	1.150	L.F.	.038	1.05	1.31	2.36	
Ceiling joists, 2" x 6", 24" O.C.	.750	L.F.	.010	.49	.34	.83	
Fascia board, 2" x 8"	.220	L.F.	.012	.15	.42	.57	
Soffit nailer (outrigger), 2" x 4", 24" O.C.	.220	L.F.	.006	.10	.20	.30	
Sheathing, 1/2" exterior plywood, CDX	1.570	S.F.	.018	.85	.63	1.48	
Furring strips, 1" x 3", 16" O.C.	1.000	L.F.	.023	.21	.80	1.01	
TOTAL			.110	2.95	3.82	6.77	
2" X 10", 16" O.C., 4/12 PITCH							
Hip rafters, 2" x 10", 4/12 pitch	.160	L.F.	.004	.21	.16	.37	
Jack rafters, 2" x 10", 16" O.C., 4/12 pitch	1.430	L.F.	.051	1.92	1.77	3.69	
Ceiling joists, 2" x 8", 16" O.C.	1.000	L.F.	.015	.91	.51	1.42	
Fascia board, 2" x 10"	.220	L.F.	.012	.15	.42	.57	
Soffit nailer (outrigger), 2" x 4", 24" O.C.	.220	L.F.	.006	.10	.20	.30	
Sheathing, 1/2" exterior plywood, CDX	1.570	S.F.	.018	.85	.63	1.48	
Furring strips, 1" x 3", 16" O.C.	1.000	L.F.	.023	.21	.80	1.01	
TOTAL			.129	4.35	4.49	8.84	
2" X 10", 24" O.C., 4/12 PITCH							
Hip rafters, 2" x 10", 4/12 pitch	.160	L.F.	.004	.21	.16	.37	
Jack rafters, 2" x 8", 24" O.C., 4/12 pitch	1.150	L.F.	.038	1.05	1.31	2.36	
Ceiling joists, 2" x 8", 24" O.C.	.750	L.F.	.011	.68	.38	1.06	
Fascia board, 2" x 10"	.220	L.F.	.012	.15	.42	.57	
Soffit nailer (outrigger), 2" x 4", 24" O.C.	.220	L.F.	.006	.10	.20	.30	
Sheathing, 1/2" exterior plywood, CDX	1.570	S.F.	.018	.85	.63	1.48	
Furring strips, 1" x 3", 16" O.C.	1.000	L.F.	.023	.21	.80	1.01	
TOTAL			.112	3.25	3.90	7.15	

The cost of this system is based on S.F. of plan area. Measurement is area under the hip roof only. See gable roof system for added costs.

Hip Roof Framing System Components

Component Description	QUAN.	UNIT	LABOR-HOURS	COST PER S.F. MAT.	COST PER S.F. INST.	COST PER S.F. TOTAL
Jack rafters						
Jack rafters, #2 or better, 16" O.C., 2" x 6", 4/12 pitch	1.430	L.F.	.038	.93	1.33	2.26
8/12 pitch	1.800	L.F.	.061	1.17	2.11	3.28
2" x 8", 4/12 pitch	1.430	L.F.	.047	1.30	1.63	2.93
8/12 pitch	1.800	L.F.	.075	1.64	2.61	4.25
24" O.C., 2" x 6", 4/12 pitch	1.150	L.F.	.031	.75	1.07	1.82
8/12 pitch	1.440	L.F.	.048	.94	1.68	2.62
2" x 8", 4/12 pitch	1.150	L.F.	.038	1.05	1.31	2.36
8/12 pitch	1.440	L.F.	.066	1.93	2.29	4.22
Ceiling Joists						
2" x 10", 16" O.C.	1.000	L.F.	.018	1.34	.62	1.96
24" O.C.	.750	L.F.	.013	1.01	.47	1.48
Sheathing						
Sheathing, plywood CDX, 4/12 pitch, 3/8" thick	1.570	S.F.	.016	.64	.58	1.22
5/8" thick	1.570	S.F.	.019	1.07	.68	1.75

Gambrel Roof Framing Systems

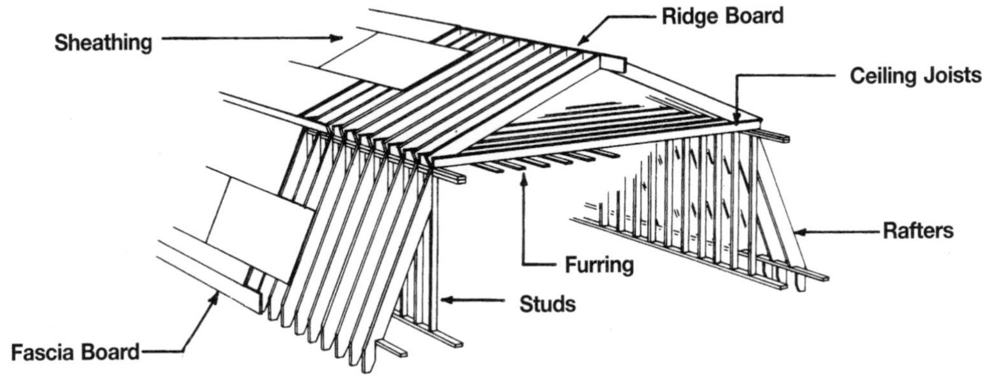

System Description	QUAN.	UNIT	LABOR-HOURS	COST PER S.F. MAT.	COST PER S.F. INST.	COST PER S.F. TOTAL	S.F. EXTENSIONS
2″ X 6″ RAFTERS, 16″ O.C.							
Roof rafters, 2″ x 6″, 16″ O.C.	1.430	L.F.	.029	.93	1.00	1.93	
Ceiling joists, 2″ x 6″, 16″ O.C.	.710	L.F.	.009	.46	.32	.78	
Stud wall, 2″ x 4″, 16″ O.C., including plates	.790	L.F.	.012	.35	.43	.78	
Furring strips, 1″ x 3″, 16″ O.C.	.710	L.F.	.016	.15	.57	.72	
Ridge board, 2″ x 8″	.050	L.F.	.002	.05	.06	.11	
Fascia board, 2″ x 6″	.100	L.F.	.006	.07	.20	.27	
Sheathing, exterior grade plywood, 1/2″ thick	1.450	S.F.	.017	.78	.58	1.36	
TOTAL			.091	2.79	3.16	5.95	
2″ X 6″ RAFTERS, 24″ O.C.							
Roof rafters, 2″ x 6″, 24″ O.C.	1.140	L.F.	.023	.74	.80	1.54	
Ceiling joists, 2″ x 6″, 24″ O.C.	.570	L.F.	.007	.37	.26	.63	
Stud wall, 2″ x 4″, 24″ O.C., including plates	.630	L.F.	.010	.28	.35	.63	
Furring strips, 1″ x 3″, 16″ O.C.	.710	L.F.	.016	.15	.57	.72	
Ridge board, 2″ x 8″	.050	L.F.	.002	.05	.06	.11	
Fascia board, 2″ x 6″	.100	L.F.	.006	.07	.20	.27	
Sheathing, exterior grade plywood, 1/2″ thick	1.450	S.F.	.017	.78	.58	1.36	
TOTAL			.081	2.44	2.82	5.26	
2″ X 8″ RAFTERS, 16″ O.C.							
Roof rafters, 2″ x 8″, 16″ O.C.	1.430	L.F.	.031	1.30	1.06	2.36	
Ceiling joists, 2″ x 6″, 16″ O.C.	.710	L.F.	.009	.46	.32	.78	
Stud wall, 2″ x 4″, 16″ O.C., including plates	.790	L.F.	.012	.35	.43	.78	
Furring strips, 1″ x 3″, 16″ O.C.	.710	L.F.	.016	.15	.57	.72	
Ridge board, 2″ x 8″	.050	L.F.	.002	.05	.06	.11	
Fascia board, 2″ x 8″	.100	L.F.	.007	.09	.25	.34	
Sheathing, exterior grade plywood, 1/2″ thick	1.450	S.F.	.017	.78	.58	1.36	
TOTAL			.094	3.18	3.27	6.45	

Gambrel Roof Framing Systems

System Description	QUAN.	UNIT	LABOR-HOURS	COST PER S.F. MAT.	COST PER S.F. INST.	COST PER S.F. TOTAL	____S.F. EXTENSIONS
2" X 8" RAFTERS, 24" O.C.							
Roof rafters, 2" x 8", 24" O.C.	1.140	L.F.	.024	1.04	.84	1.88	
Ceiling joists, 2" x 6", 24" O.C.	.570	L.F.	.007	.37	.26	.63	
Stud wall, 2" x 4", 24" O.C., including plates	.630	L.F.	.010	.28	.35	.63	
Furring strips, 1" x 3", 16" O.C.	.710	L.F.	.016	.15	.57	.72	
Ridge board, 2" x 8"	.050	L.F.	.002	.05	.06	.11	
Fascia board, 2" x 8"	.100	L.F.	.007	.09	.25	.34	
Sheathing, exterior grade plywood, 1/2" thick	1.450	S.F.	.017	.78	.58	1.36	
TOTAL			.083	2.76	2.91	5.67	
2" X 10" RAFTERS, 16" O.C.							
Roof rafters, 2" x 10", 16" O.C.	1.430	L.F.	.046	1.92	1.60	3.52	
Ceiling joists, 2" x 8", 16" O.C.	.710	L.F.	.010	.65	.36	1.01	
Stud wall, 2" x 6", 16" O.C., including plates	.790	L.F.	.014	.51	.50	1.01	
Furring strips, 1" x 3", 16" O.C.	.710	L.F.	.016	.15	.57	.72	
Ridge board, 2" x 10"	.050	L.F.	.002	.07	.07	.14	
Fascia board, 2" x 10"	.100	L.F.	.009	.13	.31	.44	
Sheathing, exterior grade plywood, 1/2" thick	1.450	S.F.	.017	.78	.58	1.36	
TOTAL			.114	4.21	3.99	8.20	
2" X 10" RAFTERS, 24" O.C.							
Roof rafters, 2" x 10", 24" O.C.	1.140	L.F.	.037	1.53	1.28	2.81	
Ceiling joists, 2" x 8", 24" O.C.	.570	L.F.	.008	.52	.29	.81	
Stud wall, 2" x 6", 24" O.C., including plates	.630	L.F.	.011	.41	.40	.81	
Furring strips, 1" x 3", 16" O.C.	.710	L.F.	.016	.15	.57	.72	
Ridge board, 2" x 10"	.050	L.F.	.002	.07	.07	.14	
Fascia board, 2" x 10"	.100	L.F.	.009	.13	.31	.44	
Sheathing, exterior grade plywood, 1/2" thick	1.450	S.F.	.017	.78	.58	1.36	
TOTAL			.100	3.59	3.50	7.09	

The cost of this system is based on the square foot of plan area on the first floor.

Gambrel Roof Framing System Components

Component Description	QUAN.	UNIT	LABOR-HOURS	COST PER S.F. MAT.	COST PER S.F. INST.	COST PER S.F. TOTAL
Ceiling joists						
Ceiling joist, #2 or better, 2" x 4", 16" O.C.	.710	L.F.	.009	.31	.32	.63
24" O.C.	.570	L.F.	.007	.25	.26	.51
Furring						
Furring, 1" x 3", 16" O.C.	.710	L.F.	.016	.15	.57	.72
24" O.C.	.590	L.F.	.013	.12	.47	.59
Ridge board						
Ridge board, #2 or better, 1" x 6"	.050	L.F.	.001	.05	.05	.10
1" x 8"	.050	L.F.	.001	.07	.05	.12
1" x 10"	.050	L.F.	.002	.08	.06	.14
2" x 6"	.050	L.F.	.002	.03	.06	.09
Fascia board						
Fascia board, #2 or better, 1" x 6"	.100	L.F.	.004	.05	.14	.19
1" x 8"	.100	L.F.	.005	.06	.16	.22
1" x 10"	.100	L.F.	.005	.07	.18	.25

Mansard Roof Framing Systems

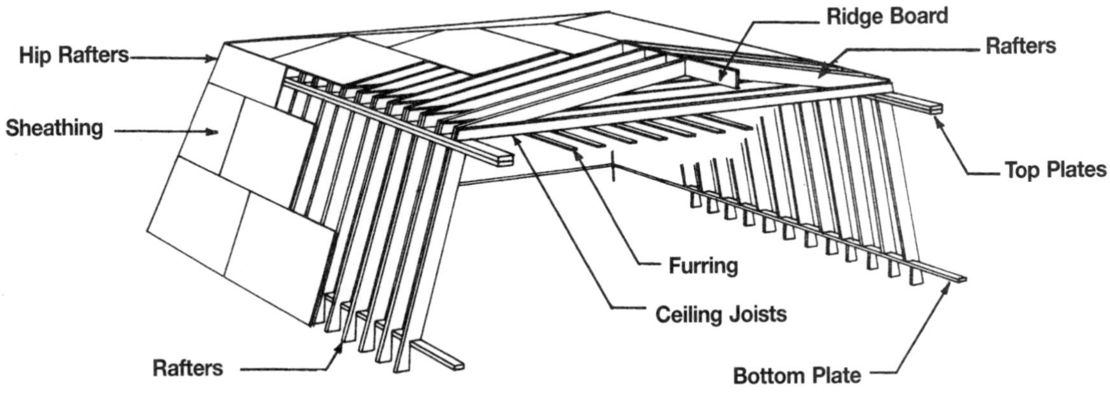

System Description	QUAN.	UNIT	LABOR-HOURS	COST PER S.F.			____S.F. EXTENSIONS
				MAT.	INST.	TOTAL	
2″ X 6″ RAFTERS, 16″ O.C.							
Roof rafters, 2″ x 6″, 16″ O.C.	1.210	L.F.	.033	.79	1.14	1.93	
Rafter plates, 2″ x 6″, double top, single bottom	.364	L.F.	.010	.24	.34	.58	
Ceiling joists, 2″ x 4″, 16″ O.C.	.920	L.F.	.012	.40	.41	.81	
Hip rafter, 2″ x 6″	.070	L.F.	.002	.05	.08	.13	
Jack rafter, 2″ x 6″, 16″ O.C.	1.000	L.F.	.039	.65	1.36	2.01	
Ridge board, 2″ x 6″	.018	L.F.	.001	.01	.02	.03	
Sheathing, exterior grade plywood, 1/2″ thick	2.210	S.F.	.025	1.19	.88	2.07	
Furring strips, 1″ x 3″, 16″ O.C.	.920	L.F.	.021	.19	.74	.93	
TOTAL			.143	3.52	4.97	8.49	
2″ X 6″ RAFTERS, 24″ O.C.							
Roof rafters, 2″ x 6″, 24″ O.C.	.970	L.F.	.026	.63	.91	1.54	
Rafter plates, 2″ x 6″, double top, single bottom	.364	L.F.	.010	.24	.34	.58	
Ceiling joists, 2″ x 4″, 24″ O.C.	.740	L.F.	.009	.33	.33	.66	
Hip rafter, 2″ x 6″	.070	L.F.	.002	.05	.08	.13	
Jack rafter, 2″ x 6″, 24″ O.C.	.800	L.F.	.031	.52	1.09	1.61	
Ridge board, 2″ x 6″	.018	L.F.	.001	.01	.02	.03	
Sheathing, exterior grade plywood, 1/2″ thick	2.210	S.F.	.025	1.19	.88	2.07	
Furring strips, 1″ x 3″, 16″ O.C.	.920	L.F.	.021	.19	.74	.93	
TOTAL			.125	3.16	4.39	7.55	
2″ X 8″ RAFTERS, 16″ O.C.							
Roof rafters, 2″ x 8″, 16″ O.C.	1.210	L.F.	.036	1.10	1.25	2.35	
Rafter plates, 2″ x 8″, double top, single bottom	.364	L.F.	.011	.33	.37	.70	
Ceiling joists, 2″ x 6″, 16″ O.C.	.920	L.F.	.012	.60	.41	1.01	
Hip rafter, 2″ x 8″	.070	L.F.	.002	.06	.08	.14	
Jack rafter, 2″ x 8″, 16″ O.C.	1.000	L.F.	.048	.91	1.66	2.57	
Ridge board, 2″ x 8″	.018	L.F.	.001	.02	.02	.04	
Sheathing, exterior grade plywood, 1/2″ thick	2.210	S.F.	.025	1.19	.88	2.07	
Furring strips, 1″ x 3″, 16″ O.C.	.920	L.F.	.021	.19	.74	.93	
TOTAL			.156	4.40	5.41	9.81	

Mansard Roof Framing Systems

System Description	QUAN.	UNIT	LABOR-HOURS	COST PER S.F. MAT.	COST PER S.F. INST.	COST PER S.F. TOTAL	____S.F. EXTENSIONS
2" X 8" RAFTERS, 24" O.C.							
Roof rafters, 2" x 8", 24" O.C.	.970	L.F.	.029	.88	1.00	1.88	
Rafter plates, 2" x 8", double top, single bottom	.364	L.F.	.011	.33	.37	.70	
Ceiling joists, 2" x 6", 24" O.C.	.740	L.F.	.011	.67	.38	1.05	
Hip rafter, 2" x 8"	.070	L.F.	.002	.06	.08	.14	
Jack rafter, 2" x 8", 24" O.C.	.800	L.F.	.038	.73	1.33	2.06	
Ridge board, 2" x 8"	.018	L.F.	.001	.02	.02	.04	
Sheathing, exterior grade plywood, 1/2" thick	2.210	S.F.	.025	1.19	.88	2.07	
Furring strips, 1" x 3", 16" O.C.	.920	L.F.	.021	.19	.74	.93	
TOTAL			.138	4.07	4.80	8.87	
2" X 10" RAFTERS, 16" O.C.							
Roof rafters, 2" x 10", 16" O.C.	1.210	L.F.	.046	1.62	1.59	3.21	
Rafter plates, 2" x 10", double top, single bottom	.364	L.F.	.014	.49	.48	.97	
Ceiling joists, 2" x 8", 16" O.C.	.920	L.F.	.013	.84	.47	1.31	
Hip rafter, 2" x 10"	.070	L.F.	.003	.09	.10	.19	
Jack rafter, 2" x 8", 16" O.C.	1.000	L.F.	.048	.91	1.66	2.57	
Ridge board, 2" x 10"	.018	L.F.	.001	.02	.03	.05	
Sheathing, exterior grade plywood, 1/2" thick	2.210	S.F.	.025	1.19	.88	2.07	
Furring strips, 1" x 3", 16" O.C.	.920	L.F.	.021	.19	.74	.93	
TOTAL			.171	5.35	5.95	11.30	
2" X 10" RAFTERS, 24" O.C.							
Roof rafters, 2" x 10", 24" O.C.	.970	L.F.	.037	1.30	1.27	2.57	
Rafter plates, 2" x 10", double top, single bottom	.364	L.F.	.014	.49	.48	.97	
Ceiling joists, 2" x 8", 24" O.C.	.740	L.F.	.011	.67	.38	1.05	
Hip rafter, 2" x 10"	.070	L.F.	.003	.09	.10	.19	
Jack rafter, 2" x 8", 24" O.C.	.800	L.F.	.038	.73	1.33	2.06	
Ridge board, 2" x 10"	.018	L.F.	.001	.02	.03	.05	
Sheathing, exterior grade plywood, 1/2" thick	2.210	S.F.	.025	1.19	.88	2.07	
Furring strips, 1" x 3", 16" O.C.	.920	L.F.	.021	.19	.74	.93	
TOTAL			.150	4.68	5.21	9.89	

The cost of this system is based on the square foot of plan area.

Mansard Roof Framing System Components

Component Description	QUAN.	UNIT	LABOR-HOURS	COST PER S.F. MAT.	COST PER S.F. INST.	COST PER S.F. TOTAL
Ridge board, #2 or better						
Ridge board, #2 or better, 1" x 6"	.018	L.F.	.001	.02	.02	.04
1" x 8"	.018	L.F.	.001	.02	.02	.04
Sheathing						
Sheathing, plywood exterior grade CDX, 3/8" thick	2.210	S.F.	.023	.91	.82	1.73
1/2" thick	2.210	S.F.	.025	1.19	.88	2.07
5/8" thick	2.210	S.F.	.027	1.50	.95	2.45
3/4" thick	2.210	S.F.	.029	1.81	1.02	2.83
Boards, 1" x 6", laid regular	2.210	S.F.	.049	3.16	1.70	4.86
Laid diagonal	2.210	S.F.	.054	3.16	1.90	5.06
1" x 8", laid regular	2.210	S.F.	.040	3.16	1.41	4.57
Laid diagonal	2.210	S.F.	.049	3.16	1.70	4.86

Shed/Flat Roof Framing Systems

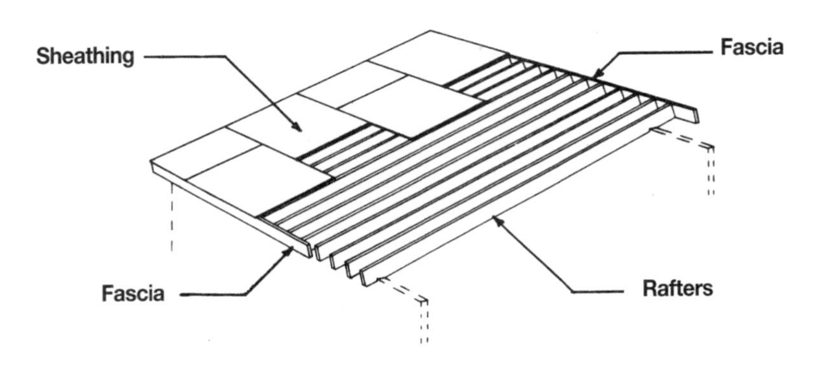

System Description	QUAN.	UNIT	LABOR-HOURS	COST PER S.F. MAT.	COST PER S.F. INST.	COST PER S.F. TOTAL	____S.F. EXTENSIONS
2″ X 4″,16″ O.C., 4/12 PITCH							
Rafters, 2″ x 4″, 16″ O.C., 4/12 pitch	1.170	L.F.	.014	.57	.49	1.06	
Fascia, 2″ x 4″	.100	L.F.	.005	.06	.17	.23	
Bridging, 1″ x 3″, 6′ OC	.080	Pr.	.005	.02	.17	.19	
Sheathing, exterior grade plywood, 1/2″ thick	1.230	S.F.	.014	.66	.49	1.15	
TOTAL			.038	1.31	1.32	2.63	
2″ X 4″, 24″ O.C., 4/12 PITCH							
Rafters, 2″ x 4″, 24″ O.C., 4/12 pitch	.940	L.F.	.011	.46	.40	.86	
Fascia, 2″ x 4″	.100	L.F.	.005	.06	.17	.23	
Bridging, 1″ x 3″, 6′ OC	.060	Pr.	.005	.02	.17	.19	
Sheathing, exterior grade plywood, 1/2″ thick	1.230	S.F.	.014	.66	.49	1.15	
TOTAL			.035	1.20	1.23	2.43	
2″ X 6″,16″ O.C., 4/12 PITCH							
Rafters, 2″ x 6″, 16″ O.C., 4/12 pitch	1.170	L.F.	.019	.76	.66	1.42	
Fascia, 2″ x 6″	.100	L.F.	.006	.07	.20	.27	
Bridging, 1″ x 3″, 6′ O.C.	.080	Pr.	.005	.02	.17	.19	
Sheathing, exterior grade plywood, 1/2″ thick	1.230	S.F.	.014	.66	.49	1.15	
TOTAL			.044	1.51	1.52	3.03	
2″ X 6″, 24″ O.C., 4/12 PITCH							
Rafters, 2″ x 6″, 24″ O.C., 4/12 pitch	.940	L.F.	.015	.61	.53	1.14	
Fascia, 2″ x 6″	.100	L.F.	.006	.07	.20	.27	
Bridging, 1″ x 3″, 6′ O.C.	.060	Pr.	.004	.02	.13	.15	
Sheathing, exterior grade plywood, 1/2″ thick	1.230	S.F.	.014	.66	.49	1.15	
TOTAL			.039	1.36	1.35	2.71	
2″ X 8″, 16″ O.C., 4/12 PITCH							
Rafters, 2″ x 8″, 16″ O.C., 4/12 pitch	1.170	L.F.	.020	1.06	.69	1.75	
Fascia, 2″ x 8″	.100	L.F.	.007	.09	.25	.34	
Bridging, 1″ x 3″, 6′ O.C.	.080	Pr.	.005	.02	.17	.19	
Sheathing, exterior grade plywood, 1/2″ thick	1.230	S.F.	.014	.66	.49	1.15	
TOTAL			.046	1.83	1.60	3.43	

Shed/Flat Roof Framing Systems

System Description	QUAN.	UNIT	LABOR-HOURS	COST PER S.F. MAT.	INST.	TOTAL	S.F. EXTENSIONS
2" X 8", 24" O.C., 4/12 PITCH							
Rafters, 2" x 8", 24" O.C., 4/12 pitch	.940	L.F.	.016	.86	.55	1.41	
Fascia, 2" x 8"	.100	L.F.	.007	.09	.25	.34	
Bridging, 1" x 3", 6' O.C.	.060	Pr.	.004	.02	.13	.15	
Sheathing, exterior grade plywood, 1/2" thick	1.230	S.F.	.014	.66	.49	1.15	
TOTAL			.041	1.63	1.42	3.05	
2" X 10", 16" O.C., 4/12 PITCH							
Rafters, 2" x 10", 16" O.C., 4/12 pitch	1.170	L.F.	.030	1.57	1.03	2.60	
Fascia, 2" x 10"	.100	L.F.	.009	.13	.31	.44	
Bridging, 1" x 3", 6' OC	.080	Pr.	.004	.02	.13	.15	
Sheathing, exterior grade plywood, 1/2" thick	1.230	S.F.	.014	.66	.49	1.15	
TOTAL			.057	2.38	1.96	4.34	
2" X 10", 24" O.C., 4/12 PITCH							
Rafters, 2" x 10", 24" O.C., 4/12 pitch	.940	L.F.	.024	1.26	.83	2.09	
Fascia, 2" x 10"	.100	L.F.	.009	.13	.31	.44	
Bridging, 1" x 3", 6' OC	.060	Pr.	.004	.02	.13	.15	
Sheathing, exterior grade plywood, 1/2" thick	1.230	S.F.	.014	.66	.49	1.15	
TOTAL			.051	2.07	1.76	3.83	

The cost of this system is based on the square foot of plan area.
A 1' overhang is assumed. No ceiling joists or furring are included.

Shed/Flat Roof Framing System Components

Component Description	QUAN.	UNIT	LABOR-HOURS	COST PER S.F. MAT.	INST.	TOTAL
Rafters, #2 or better, 5/12 - 8/12 pitch						
Rafters, #2 or better, 16" O.C., 2" x 4", 0 - 4/12 pitch	1.170	L.F.	.014	.57	.49	1.06
2" x 8"	1.330	L.F.	.028	1.21	.98	2.19
24" O.C., 2" x 4"	1.060	L.F.	.021	.69	.74	1.43
2" x 6"	1.060	L.F.	.021	.69	.74	1.43
2" x 8"	1.060	L.F.	.023	.96	.78	1.74
2" x 10"	1.060	L.F.	.034	1.42	1.19	2.61
Fascia						
Fascia, #2 or better,, 1" x 4"	.100	L.F.	.003	.04	.10	.14
1" x 6"	.100	L.F.	.004	.05	.14	.19
Bridging						
Metal, galvanized, rafters, 16" O.C.	.080	Pr.	.005	.05	.17	.22
24" O.C.	.060	Pr.	.003	.15	.12	.27
Compression type, rafters, 16" O.C.	.080	Pr.	.003	.11	.11	.22
24" O.C.	.060	Pr.	.002	.08	.08	.16
Sheathing						
Sheathing, plywood, exterior grade, 3/8" thick, flat 0 - 4/12 pitch	1.230	S.F.	.013	.50	.46	.96
5/12 - 8/12 pitch	1.330	S.F.	.014	.55	.49	1.04
1/2" thick, flat 0 - 4/12 pitch	1.230	S.F.	.014	.66	.49	1.15
5/12 - 8/12 pitch	1.330	S.F.	.015	.72	.53	1.25
5/8" thick, flat 0 - 4/12 pitch	1.230	S.F.	.015	.84	.53	1.37
5/12 - 8/12 pitch	1.330	S.F.	.016	.90	.57	1.47

Gable Dormer Framing Systems

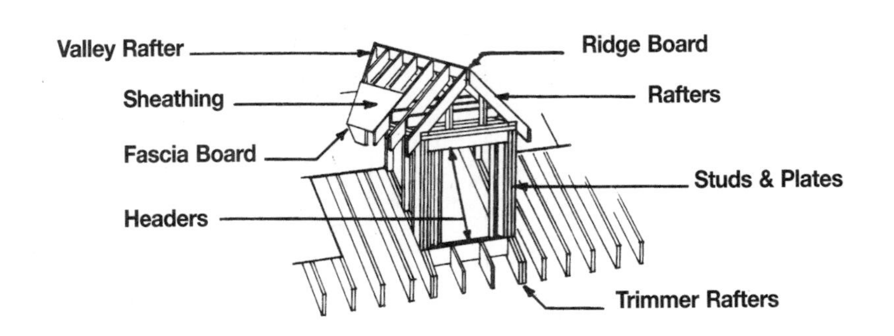

System Description	QUAN.	UNIT	LABOR-HOURS	COST PER S.F. MAT.	COST PER S.F. INST.	COST PER S.F. TOTAL	S.F. EXTENSIONS
2″ X 4″, 16″ O.C.							
Dormer rafter, 2″ x 4″, 16″ O.C.	1.330	L.F.	.029	.69	1.00	1.69	
Ridge board, 2″ x 4″	.280	L.F.	.007	.15	.25	.40	
Trimmer rafters, 2″ x 6″	.880	L.F.	.014	.57	.49	1.06	
Wall studs & plates, 2″ x 4″, 16″ O.C.	3.160	L.F.	.056	1.39	1.96	3.35	
Fascia, 2″ x 4″	.220	L.F.	.011	.14	.37	.51	
Valley rafter, 2″ x 4″, 16″ O.C.	.280	L.F.	.007	.15	.24	.39	
Cripple rafter, 2″ x 4″, 16″ O.C.	.560	L.F.	.018	.29	.61	.90	
Headers, 2″ x 4″, doubled	.670	L.F.	.024	.35	.84	1.19	
Ceiling joist, 2″ x 4″, 16″ O.C.	1.000	L.F.	.013	.44	.45	.89	
Sheathing, exterior grade plywood, 1/2″ thick	3.610	S.F.	.041	1.95	1.44	3.39	
TOTAL			.220	6.12	7.65	13.77	
2″ X 6″, 16″ O.C.							
Dormer rafter, 2″ x 6″, 16″ O.C.	1.330	L.F.	.036	.86	1.25	2.11	
Ridge board, 2″ x 6″	.280	L.F.	.009	.18	.31	.49	
Trimmer rafters, 2″ x 6″	.880	L.F.	.014	.57	.49	1.06	
Wall studs & plates, 2″ x 4″, 16″ O.C.	3.160	L.F.	.056	1.39	1.96	3.35	
Fascia, 2″ x 6″	.220	L.F.	.012	.15	.42	.57	
Valley rafter, 2″ x 6″, 16″ O.C.	.280	L.F.	.009	.18	.31	.49	
Cripple rafter, 2″ x 6″, 16″ O.C.	.560	L.F.	.022	.36	.76	1.12	
Headers, 2″ x 6″, doubled	.670	L.F.	.030	.44	1.04	1.48	
Ceiling joist, 2″ x 4″, 16″ O.C.	1.000	L.F.	.013	.44	.45	.89	
Sheathing, exterior grade plywood, 1/2″ thick	3.610	S.F.	.041	1.95	1.44	3.39	
TOTAL			.242	6.52	8.43	14.95	
2″ X 6″, 24″ O.C.							
Dormer rafter, 2″ x 6″, 24″ O.C.	1.060	L.F.	.029	.69	1.00	1.69	
Ridge board, 2″ x 6″	.280	L.F.	.009	.18	.31	.49	
Trimmer rafters, 2″ x 6″	.880	L.F.	.014	.57	.49	1.06	
Wall studs & plates, 2″ x 4″, 24″ O.C.	2.800	L.F.	.050	1.23	1.74	2.97	
Fascia, 2″ x 6″	.220	L.F.	.012	.15	.42	.57	
Valley rafter, 2″ x 6″, 24″ O.C.	.280	L.F.	.009	.18	.31	.49	
Cripple rafter, 2″ x 6″, 24″ O.C.	.450	L.F.	.018	.29	.61	.90	
Headers, 2″ x 6″, doubled	.670	L.F.	.030	.44	1.04	1.48	
Ceiling joist, 2″ x 4″, 24″ O.C.	.800	L.F.	.010	.35	.36	.71	
Sheathing, exterior grade plywood, 1/2″ thick	3.610	S.F.	.041	1.95	1.44	3.39	
TOTAL			.222	6.03	7.72	13.75	

Gable Dormer Framing Systems

System Description	QUAN.	UNIT	LABOR-HOURS	COST PER S.F. MAT.	INST.	TOTAL	___S.F. EXTENSIONS
2" X 8", 16" O.C.							
Dormer rafter, 2" x 8", 16" O.C.	1.330	L.F.	.039	1.21	1.37	2.58	
Ridge board, 2" x 8"	.280	L.F.	.010	.25	.35	.60	
Trimmer rafter, 2" x 8"	.880	L.F.	.015	.80	.52	1.32	
Wall studs & plates, 2" x 4", 16" O.C.	3.160	L.F.	.056	1.39	1.96	3.35	
Fascia, 2" x 8"	.220	L.F.	.016	.20	.54	.74	
Valley rafter, 2" x 8", 16" O.C.	.280	L.F.	.010	.25	.33	.58	
Cripple rafter, 2" x 8", 16" O.C.	.560	L.F.	.027	.51	.93	1.44	
Headers, 2" x 8", doubled	.670	L.F.	.032	.61	1.10	1.71	
Ceiling joist, 2" x 4", 16" O.C.	1.000	L.F.	.013	.44	.45	.89	
Sheathing,, exterior grade plywood, 1/2" thick	3.610	S.F.	.041	1.95	1.44	3.39	
TOTAL			.259	7.61	8.99	16.60	
2" X 8", 24" O.C.							
Dormer rafter, 2" x 8", 24" O.C.	1.060	L.F.	.031	.96	1.09	2.05	
Ridge board, 2" x 8"	.280	L.F.	.010	.25	.35	.60	
Trimmer rafter, 2" x 8"	.880	L.F.	.015	.80	.52	1.32	
Wall studs & plates, 2" x 6", 24" O.C.	2.800	L.F.	.056	1.82	1.96	3.78	
Fascia, 2" x 8"	.220	L.F.	.016	.20	.54	.74	
Valley rafter, 2" x 8", 24" O.C.	.280	L.F.	.010	.25	.33	.58	
Cripple rafter, 2" x 8", 24" O.C.	.450	L.F.	.021	.41	.75	1.16	
Headers, 2" x 8", doubled	.670	L.F.	.032	.61	1.10	1.71	
Ceiling joist, 2" x 6", 24" O.C.	.800	L.F.	.010	.52	.36	.88	
Sheathing, exterior grade plywood, 1/2" thick	3.610	S.F.	.041	1.95	1.44	3.39	
TOTAL			.242	7.77	8.44	16.21	

The cost in this system is based on the square foot of plan area.
The measurement being the plan area of the dormer only.

Gable Dormer Framing System Components

Component Description	QUAN.	UNIT	LABOR-HOURS	COST PER S.F. MAT.	INST.	TOTAL
Ridge board						
Ridge board, #2 or better, 1" x 4"	.280	L.F.	.006	.22	.21	.43
1" x 6"	.280	L.F.	.007	.28	.26	.54
1" x 8"	.280	L.F.	.008	.37	.28	.65
2" x 4"	.280	L.F.	.007	.15	.25	.40
2" x 6"	.280	L.F.	.009	.18	.31	.49
2" x 8"	.280	L.F.	.010	.25	.35	.60
Fascia						
Fascia, #2 or better, 1" x 4"	.220	L.F.	.006	.08	.22	.30
1" x 6"	.220	L.F.	.008	.10	.27	.37
1" x 8"	.220	L.F.	.009	.12	.32	.44
2" x 4"	.220	L.F.	.011	.14	.37	.51
Sheathing						
Sheathing, plywood exterior grade, 3/8" thick	3.610	S.F.	.038	1.48	1.34	2.82
5/8" thick	3.610	S.F.	.044	2.45	1.55	4.00
3/4" thick	3.610	S.F.	.048	2.96	1.66	4.62
Boards, 1" x 6", laid regular	3.610	S.F.	.089	5.15	3.10	8.25
Laid diagonal	3.610	S.F.	.099	5.15	3.43	8.58
1" x 8", laid regular	3.610	S.F.	.076	5.15	2.64	7.79

Shed Dormer Framing Systems

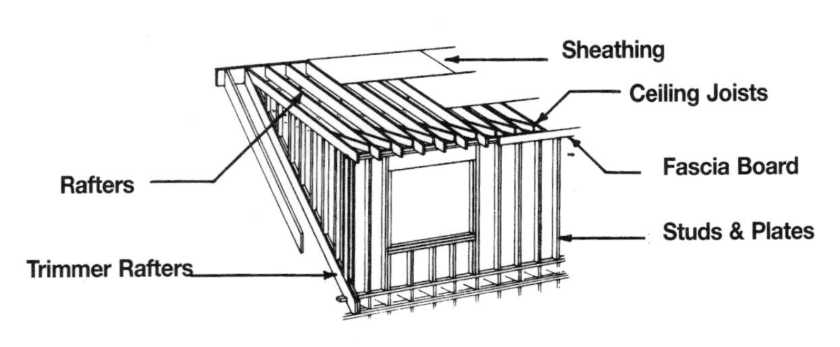

System Description	QUAN.	UNIT	LABOR-HOURS	COST PER S.F. MAT.	COST PER S.F. INST.	COST PER S.F. TOTAL	S.F. EXTENSIONS
2″ X 4″ RAFTERS, 16″ O.C.							
Dormer rafter, 2″ x 4″, 16″ O.C.	1.080	L.F.	.023	.56	.81	1.37	
Trimmer rafter, 2″ x 4″	.400	L.F.	.005	.21	.18	.39	
Studs & plates, 2″ x 4″, 16″ O.C.	2.750	L.F.	.049	1.21	1.71	2.92	
Fascia, 2″ x 4″	.250	L.F.	.011	.14	.37	.51	
Ceiling joist, 2″ x 4″, 16″ O.C.	1.000	L.F.	.013	.44	.45	.89	
Sheathing, exterior grade plywood, CDX, 1/2″ thick	2.940	S.F.	.034	1.59	1.18	2.77	
TOTAL			.135	4.15	4.70	8.85	
2″ X 6″ RAFTERS, 16″ O.C.							
Dormer rafter, 2″ x 6″, 16″ O.C.	1.080	L.F.	.029	.70	1.02	1.72	
Trimmer rafter, 2″ x 6″	.400	L.F.	.006	.26	.22	.48	
Studs & plates, 2″ x 4″, 16″ O.C.	2.750	L.F.	.049	1.21	1.71	2.92	
Fascia, 2″ x 6″	.250	L.F.	.014	.17	.47	.64	
Ceiling joist, 2″ x 4″, 16″ O.C.	1.000	L.F.	.013	.44	.45	.89	
Sheathing, exterior grade plywood, CDX, 1/2″ thick	2.940	S.F.	.034	1.59	1.18	2.77	
TOTAL			.145	4.37	5.05	9.42	
2″ X 6″ RAFTERS, 24″ O.C.							
Dormer rafter, 2″ x 6″, 24″ O.C.	.860	L.F.	.023	.56	.81	1.37	
Trimmer rafter, 2″ x 6″	.400	L.F.	.006	.26	.22	.48	
Studs & plates, 2″ x 4″, 24″ O.C.	2.750	L.F.	.049	1.21	1.71	2.92	
Fascia, 2″ x 6″	.250	L.F.	.014	.17	.47	.64	
Ceiling joist, 2″ x 4″, 24″ O.C.	.800	L.F.	.010	.35	.36	.71	
Sheathing, exterior grade plywood, CDX, 1/2″ thick	2.940	S.F.	.034	1.59	1.18	2.77	
TOTAL			.136	4.14	4.75	8.89	
2″ X 8″ RAFTERS, 16″ O.C.							
Dormer rafter, 2″ x 8″, 16″ O.C.	1.080	L.F.	.032	.98	1.11	2.09	
Trimmer rafter, 2″ x 8″	.400	L.F.	.007	.36	.24	.60	
Studs & plates, 2″ x 4″, 16″ O.C.	2.750	L.F.	.049	1.21	1.71	2.92	
Fascia, 2″ x 8″	.250	L.F.	.018	.23	.62	.85	
Ceiling joist, 2″ x 6″, 16″ O.C.	1.000	L.F.	.013	.65	.45	1.10	
Sheathing, exterior grade plywood, CDX, 1/2″ thick	2.940	S.F.	.034	1.59	1.18	2.77	
TOTAL			.153	5.02	5.31	10.33	

Shed Dormer Framing Systems

System Description	QUAN.	UNIT	LABOR-HOURS	COST PER S.F. MAT.	INST.	TOTAL	____S.F. EXTENSIONS
2" X 8" RAFTERS, 24" O.C.							
Dormer rafter, 2" x 8", 24" O.C.	.860	L.F.	.025	.78	.89	1.67	
Trimmer rafter, 2" x 8"	.400	L.F.	.007	.36	.24	.60	
Studs & plates, 2" x 4", 24" O.C.	2.750	L.F.	.055	1.79	1.93	3.72	
Fascia, 2" x 8"	.250	L.F.	.018	.23	.62	.85	
Ceiling joist, 2" x 6", 24" O.C.	.800	L.F.	.010	.52	.36	.88	
Sheathing, exterior grade plywood, CDX, 1/2" thick	2.940	S.F.	.034	1.59	1.18	2.77	
TOTAL			.149	5.27	5.22	10.49	
2" X 10" RAFTERS, 16" O.C.							
Dormer rafter, 2" x 10", 16" O.C.	1.080	L.F.	.041	1.45	1.41	2.86	
Trimmer rafter, 2" x 10"	.400	L.F.	.010	.54	.35	.89	
Studs & plates, 2" x 4", 16" O.C.	2.750	L.F.	.049	1.21	1.71	2.92	
Fascia, 2" x 10"	.250	L.F.	.022	.34	.77	1.11	
Ceiling joist, 2" x 6", 16" O.C.	1.000	L.F.	.013	.65	.45	1.10	
Sheathing, exterior grade plywood, CDX, 1/2" thick	2.940	S.F.	.034	1.59	1.18	2.77	
TOTAL			.169	5.78	5.87	11.65	
2" X 10" RAFTERS, 24" O.C.							
Dormer rafter, 2" x 10", 24" O.C.	.860	L.F.	.032	1.15	1.13	2.28	
Trimmer rafter, 2" x 10"	.400	L.F.	.010	.54	.35	.89	
Studs & plates, 2" x 4", 24" O.C.	2.750	L.F.	.055	1.79	1.93	3.72	
Fascia, 2" x 10"	.250	L.F.	.022	.34	.77	1.11	
Ceiling joist, 2" x 6", 24" O.C.	.800	L.F.	.010	.52	.36	.88	
Sheathing, exterior grade plywood, CDX, 1/2" thick	2.940	S.F.	.034	1.59	1.18	2.77	
TOTAL			.163	5.93	5.72	11.65	

The cost in this system is based on the square foot of plan area.
The measurement is the plan area of the dormer only.

Shed Dormer Framing System Components

Component Description	QUAN.	UNIT	LABOR-HOURS	COST PER S.F. MAT.	INST.	TOTAL
Fascia boards						
Fascia, #2 or better, 1" x 4"	.250	L.F.	.006	.08	.22	.30
1" x 6"	.250	L.F.	.008	.10	.27	.37
Ceiling joists						
2" x 8", 16" O.C.	1.000	L.F.	.015	.91	.51	1.42
24" O.C.	.800	L.F.	.012	.73	.41	1.14
Sheathing						
Sheathing, plywood exterior grade, 3/8" thick	2.940	S.F.	.031	1.21	1.09	2.30
5/8" thick	2.940	S.F.	.036	2.00	1.26	3.26

Window Openings

Component Description	QUAN.	UNIT	LABOR-HOURS	COST EACH MAT.	INST.	TOTAL
The following are to be added to the total cost of the dormers for window openings. Do not subtract window area from the stud wall quantities.						
Headers						
2" x 6" doubled, 2' long	4.000	L.F.	.178	2.60	6.20	8.80
4' long	8.000	L.F.	.356	5.20	12.40	17.60
2" x 8" doubled, 4' long	8.000	L.F.	.376	7.30	13.10	20.40
8' long	16.000	L.F.	.753	14.55	26.00	40.55

Partition Framing Systems

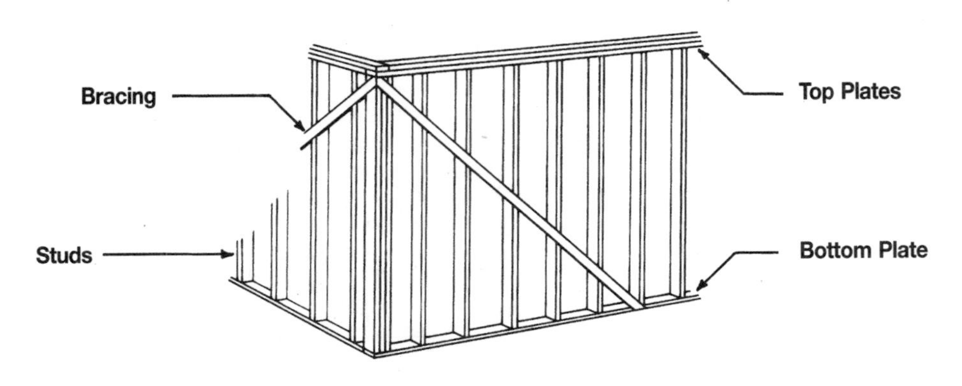

System Description	QUAN.	UNIT	LABOR-HOURS	COST PER S.F. MAT.	COST PER S.F. INST.	COST PER S.F. TOTAL	____S.F. EXTENSIONS
2″ X 4″, 12″ O.C.							
2″ x 4″ studs, #2 or better, 12″ O.C.	1.250	L.F.	.018	.55	.64	1.19	
Plates, double top, single bottom	.375	L.F.	.005	.17	.19	.36	
Cross bracing, let-in, 1″ x 6″	.080	L.F.	.004	.05	.15	.20	
TOTAL			.027	.77	.98	1.75	
2″ X 4″, 16″ O.C.							
2″ x 4″ studs, #2 or better, 16″ O.C.	1.000	L.F.	.015	.44	.51	.95	
Plates, double top, single bottom	.375	L.F.	.005	.17	.19	.36	
Cross bracing, let-in, 1″ x 6″	.080	L.F.	.004	.05	.15	.20	
TOTAL			.024	.66	.85	1.51	
2″ X 4″, 24″ O.C.							
2″ x 4″ studs, #2 or better, 24″ O.C.	.800	L.F.	.012	.35	.41	.76	
Plates, double top, single bottom	.375	L.F.	.005	.17	.19	.36	
Cross bracing, let-in, 1″ x 6″	.080	L.F.	.003	.05	.10	.15	
TOTAL			.020	.57	.70	1.27	
2″ X 4″, 32″ O.C.							
2″ x 4″ studs, #2 or better, 32″ O.C.	.650	L.F.	.009	.29	.33	.62	
Plates, double top, single bottom	.375	L.F.	.005	.17	.19	.36	
Cross bracing, let-in, 1″ x 6″	.080	L.F.	.003	.05	.10	.15	
TOTAL			.017	.51	.62	1.13	
2″ X 6″, 12″ O.C.							
2″ x 6″ studs, #2 or better, 12″ O.C.	1.250	L.F.	.020	.81	.70	1.51	
Plates, double top, single bottom	.375	L.F.	.006	.24	.21	.45	
Cross bracing, let-in, 1″ x 6″	.080	L.F.	.004	.05	.15	.20	
TOTAL			.030	1.10	1.06	2.16	
2″ X 6″, 16″ O.C.							
2″ x 6″ studs, #2 or better, 16″ O.C.	1.000	L.F.	.016	.65	.56	1.21	
Plates, double top, single bottom	.375	L.F.	.006	.24	.21	.45	
Cross bracing, let-in, 1″ x 6″	.080	L.F.	.004	.05	.15	.20	
TOTAL			.026	.94	.92	1.86	

Partition Framing Systems

System Description	QUAN.	UNIT	LABOR-HOURS	COST PER S.F. MAT.	COST PER S.F. INST.	COST PER S.F. TOTAL	S.F. EXTENSIONS
2″ X 6″, 24″ O.C.							
2″ x 6″ studs, #2 or better, 24″ O.C.	.800	L.F.	.013	.52	.45	.97	
Plates, double top, single bottom	.375	L.F.	.006	.24	.21	.45	
Cross bracing, let-in, 1″ x 6″	.080	L.F.	.003	.05	.10	.15	
TOTAL			.022	.81	.76	1.57	
2″ X 6″, 32″ O.C.							
2″ x 6″ studs, #2 or better, 32″ O.C.	.650	L.F.	.010	.42	.36	.78	
Plates, double top, single bottom	.375	L.F.	.006	.24	.21	.45	
Cross bracing, let-in, 1″ x 6″	.080	L.F.	.003	.05	.10	.15	
TOTAL			.019	.71	.67	1.38	

The costs in this system are based on a square foot of wall area. Do not subtract for door or window openings.

Partition Framing System Components

Component Description	QUAN.	UNIT	LABOR-HOURS	COST PER S.F. MAT.	COST PER S.F. INST.	COST PER S.F. TOTAL
Cross bracing						
Let-in steel (T shaped) studs, 12″ O.C.	.080	L.F.	.001	.04	.05	.09
16″ O.C.	.080	L.F.	.001	.04	.04	.08
24″ O.C.	.080	L.F.	.001	.04	.04	.08
32″ O.C.	.080	L.F.	.001	.03	.03	.06
Steel straps studs, 12″ O.C.	.080	L.F.	.001	.06	.04	.10
16″ O.C.	.080	L.F.	.001	.05	.04	.09
24″ O.C.	.080	L.F.	.001	.05	.04	.09
32″ O.C.	.080	L.F.	.001	.05	.03	.08
Metal studs						
Load bearing 24″ O.C., 20 ga. galv., 2-1/2″ wide	1.000	S.F.	.015	.35	.52	.87
3-5/8″ wide	1.000	S.F.	.015	.42	.53	.95
4″ wide	1.000	S.F.	.016	.44	.54	.98
6″ wide	1.000	S.F.	.016	.56	.55	1.11
16 ga., 2-1/2″ wide	1.000	S.F.	.017	.41	.59	1.00
3-5/8″ wide	1.000	S.F.	.017	.49	.61	1.10
4″ wide	1.000	S.F.	.018	.52	.62	1.14
6″ wide	1.000	S.F.	.018	.66	.64	1.30
Non-load bearing 24″ O.C., 25 ga. galv., 1-5/8″ wide	1.000	S.F.	.011	.12	.37	.49
2-1/2″ wide	1.000	S.F.	.011	.13	.37	.50
3-5/8″ wide	1.000	S.F.	.011	.16	.38	.54
4″ wide	1.000	S.F.	.011	.19	.38	.57
6″ wide	1.000	S.F.	.011	.25	.38	.63
20 ga., 2-1/2″ wide	1.000	S.F.	.013	.22	.46	.68
3-5/8″ wide	1.000	S.F.	.014	.24	.47	.71
4″ wide	1.000	S.F.	.014	.30	.47	.77
6″ wide	1.000	S.F.	.014	.35	.48	.83

Exterior Closure Systems

Wood Siding Systems

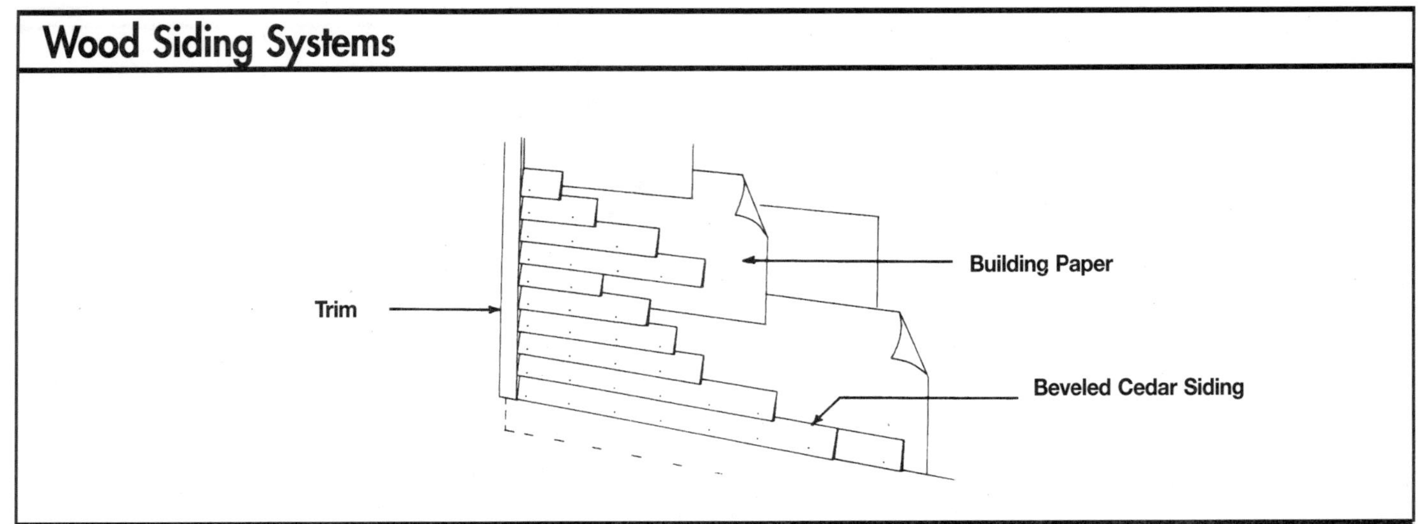

System Description	QUAN.	UNIT	LABOR-HOURS	COST PER S.F. MAT.	COST PER S.F. INST.	COST PER S.F. TOTAL	____S.F. EXTENSIONS
1/2″ X 6″ BEVELED CEDAR SIDING, "A" GRADE							
1/2″ x 6″ beveled cedar siding	1.000	S.F.	.032	1.82	1.11	2.93	
#15 asphalt felt paper	1.100	S.F.	.002	.02	.09	.11	
Trim, cedar	.125	L.F.	.005	.18	.17	.35	
Paint, primer & 2 coats	1.000	S.F.	.017	.17	.53	.70	
TOTAL			.056	2.19	1.90	4.09	
1/2″ X 8″ BEVELED CEDAR SIDING, "A" GRADE							
1/2″ x 8″ beveled cedar siding	1.000	S.F.	.029	2.09	1.01	3.10	
#15 asphalt felt paper	1.100	S.F.	.002	.02	.09	.11	
Trim, cedar	.125	L.F.	.005	.18	.17	.35	
Paint, primer & 2 coats	1.000	S.F.	.017	.17	.53	.70	
TOTAL			.053	2.46	1.80	4.26	
1″ X 10″ BOARD & BATTEN CEDAR SIDING, "B" GRADE							
1″ x 10″ board & batten cedar siding	1.000	S.F.	.031	2.22	1.07	3.29	
#15 asphalt felt paper	1.100	S.F.	.002	.02	.09	.11	
Trim, cedar	.125	L.F.	.005	.18	.17	.35	
Paint, primer & 2 coats	1.000	S.F.	.017	.17	.53	.70	
TOTAL			.055	2.59	1.86	4.45	
1″ X 10″ BOARD & BATTEN WHITE PINE SIDING							
1″ x 10″ board & batten white pine siding	1.000	S.F.	.029	.74	1.01	1.75	
#15 asphalt felt paper	1.100	S.F.	.002	.02	.09	.11	
Trim, pine	.125	L.F.	.005	.07	.17	.24	
Paint, primer & 2 coats	1.000	S.F.	.017	.17	.53	.70	
TOTAL			.053	1.00	1.80	2.80	
1″ X 4″ TONGUE & GROOVE, REDWOOD, VERTICAL GRAIN							
1″ x 4″ tongue & groove, vertical, redwood	1.000	S.F.	.033	4.40	1.16	5.56	
#15 asphalt felt paper	1.100	S.F.	.002	.02	.09	.11	
Trim, redwood	.125	L.F.	.005	.18	.17	.35	
Sealer, 1 coat, stain, 1 coat	1.000	S.F.	.013	.11	.41	.52	
TOTAL			.053	4.71	1.83	6.54	

Wood Siding Systems

System Description	QUAN.	UNIT	LABOR-HOURS	COST PER S.F. MAT.	INST.	TOTAL	S.F. EXTENSIONS
1″ X 6″ TONGUE & GROOVE, REDWOOD, VERTICAL GRAIN							
1″ x 6″ tongue & groove, vertical, redwood	1.000	S.F.	.024	4.40	.82	5.22	
#15 asphalt felt paper	1.100	S.F.	.002	.02	.09	.11	
Trim, redwood	.125	L.F.	.005	.18	.17	.35	
Sealer, 1 coat, stain, 1 coat	1.000	S.F.	.013	.11	.41	.52	
TOTAL			.044	4.71	1.49	6.20	
3/8″ PLYWOOD SIDING, TEXTURE 1-11 CEDAR							
3/8″ plywood siding	1.000	S.F.	.024	1.19	.82	2.01	
#15 asphalt felt paper	1.100	S.F.	.002	.02	.09	.11	
Trim, cedar	.125	L.F.	.005	.18	.17	.35	
Paint, primer & 2 coats	1.000	S.F.	.017	.17	.53	.70	
TOTAL			.048	1.56	1.61	3.17	
3/8″ PLYWOOD SIDING, SOUTHERN YELLOW PINE							
3/8″ plywood siding	1.000	S.F.	.024	.64	.82	1.46	
#15 asphalt felt paper	1.100	S.F.	.002	.02	.09	.11	
Trim, pine	.125	L.F.	.005	.07	.17	.24	
Paint, primer & 2 coats	1.000	S.F.	.017	.17	.53	.70	
TOTAL			.048	.90	1.61	2.51	

The costs in this system are based on a square foot of wall area.
Do not subtract area for door or window openings.

Wood Siding System Components

Component Description	QUAN.	UNIT	LABOR-HOURS	COST PER S.F. MAT.	INST.	TOTAL
Siding						
Siding, beveled cedar, "B" grade, 1/2″ x 6″	1.000	S.F.	.028	1.38	.96	2.34
1/2″ x 8″	1.000	S.F.	.023	1.38	.80	2.18
Clear grade, 1/2″ x 6″	1.000	S.F.	.028	1.49	.96	2.45
1/2″ x 8″	1.000	S.F.	.023	1.49	.80	2.29
Redwood, clear vertical grain, 1/2″ x 6″	1.000	S.F.	.036	2.88	1.24	4.12
1/2″ x 8″	1.000	S.F.	.032	2.33	1.11	3.44
Siding board & batten, cedar, "B" grade, 1″ x 10″	1.000	S.F.	.031	2.22	1.07	3.29
1″ x 12″	1.000	S.F.	.031	2.22	1.07	3.29
White pine, #2 & better, 1″ x 10″	1.000	S.F.	.029	.74	1.01	1.75
1″ x 12″	1.000	S.F.	.029	.74	1.01	1.75
Siding vertical, tongue & groove, cedar "B" grade, 1″ x 4″	1.000	S.F.	.033	3.06	1.16	4.22
1″ x 8″	1.000	S.F.	.024	2.29	.82	3.11
"A" grade, 1″ x 4″	1.000	S.F.	.033	4.55	1.16	5.71
1″ x 8″	1.000	S.F.	.024	3.43	.82	4.25
Clear vertical grain, 1″ x 4″	1.000	S.F.	.033	5.10	1.16	6.26
1″ x 6″	1.000	S.F.	.024	3.94	.82	4.76
1″ x 8″	1.000	S.F.	.024	4.13	.82	4.95
Siding plywood, texture 1-11 cedar, 3/8″ thick	1.000	S.F.	.024	1.19	.82	2.01
5/8″ thick	1.000	S.F.	.024	1.68	.82	2.50
Redwood, 3/8″ thick	1.000	S.F.	.024	1.19	.82	2.01
5/8″ thick	1.000	S.F.	.024	1.98	.82	2.80
Fir, 3/8″ thick	1.000	S.F.	.024	.64	.82	1.46
5/8″ thick	1.000	S.F.	.024	.95	.82	1.77
Hard board, 7/16″ thick primed, plain finish	1.000	S.F.	.025	1.14	.86	2.00

Shingle Siding Systems

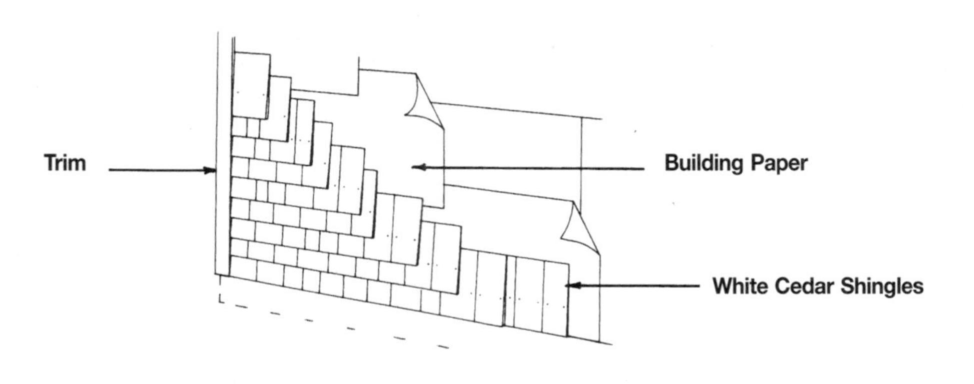

System Description	QUAN.	UNIT	LABOR-HOURS	COST PER S.F. MAT.	COST PER S.F. INST.	COST PER S.F. TOTAL	____S.F. EXTENSIONS
WHITE CEDAR SHINGLES, 5″ EXPOSURE							
White cedar shingles, 16″ long, grade "A", 5″ exposure	1.000	S.F.	.033	1.23	1.16	2.39	
#15 asphalt felt paper	1.100	S.F.	.002	.02	.09	.11	
Trim, cedar	.125	S.F.	.005	.18	.17	.35	
Paint, primer & 1 coat	1.000	S.F.	.017	.09	.54	.63	
TOTAL			.057	1.52	1.96	3.48	
WHITE CEDAR SHINGLES, 5″ EXPOSURE, GRADE "B"							
White cedar shingles, 16″ long, grade "B", 5″ exp.	1.000	S.F.	.040	1.21	1.39	2.60	
#15 asphalt felt paper	1.100	S.F.	.002	.02	.09	.11	
Trim, cedar	.125	S.F.	.005	.18	.17	.35	
Paint, primer & 1 coat	1.000	S.F.	.017	.09	.54	.63	
TOTAL			.064	1.50	2.19	3.69	
FIRE RETARDANT, WHITE CEDAR SHINGLES, 5″ EXPOSURE							
White cedar shingles, 16″ long, grade "A", 5″ exp.	1.000	S.F.	.033	1.56	1.16	2.72	
#15 asphalt felt paper	1.100	S.F.	.002	.02	.09	.11	
Trim, cedar	.125	S.F.	.005	.18	.17	.35	
Paint, primer & 1 coat	1.000	S.F.	.017	.09	.54	.63	
TOTAL			.057	1.85	1.96	3.81	
NO. 1 PERFECTIONS, 5-1/2″ EXPOSURE							
No. 1 perfections, red cedar, 5-1/2″ exposure	1.000	S.F.	.029	1.76	1.01	2.77	
#15 asphalt felt paper	1.100	S.F.	.002	.02	.09	.11	
Trim, cedar	.125	S.F.	.005	.18	.17	.35	
Stain, sealer & 1 coat	1.000	S.F.	.017	.09	.54	.63	
TOTAL			.053	2.05	1.81	3.86	
NO. 1 PERFECTIONS, 10″ EXPOSURE							
No. 1 perfections, red cedar, 10″ exposure	1.000	S.F.	.025	.90	.87	1.77	
#15 asphalt felt paper	1.100	S.F.	.002	.02	.09	.11	
Trim, cedar	.125	S.F.	.005	.18	.17	.35	
Stain, sealer & 1 coat	1.000	S.F.	.017	.09	.54	.63	
TOTAL			.049	1.19	1.67	2.86	

Shingle Siding Systems

System Description	QUAN.	UNIT	LABOR-HOURS	COST PER S.F. MAT.	COST PER S.F. INST.	COST PER S.F. TOTAL	____S.F. EXTENSIONS
RESQUARED & REBUTTED PERFECTIONS, 5-1/2″ EXPOSURE							
Resquared & rebutted perfections, 5-1/2″ exposure	1.000	S.F.	.027	2.19	.93	3.12	
#15 asphalt felt paper	1.100	S.F.	.002	.02	.09	.11	
Trim, cedar	.125	S.F.	.005	.18	.17	.35	
Stain, sealer & 1 coat	1.000	S.F.	.017	.09	.54	.63	
TOTAL			.051	2.48	1.73	4.21	
HAND-SPLIT SHAKES, 8-1/2″ EXPOSURE							
Hand-split red cedar shakes, 18″ long, 8-1/2″ exposure	1.000	S.F.	.040	1.07	1.39	2.46	
#15 asphalt felt paper	1.100	S.F.	.002	.02	.09	.11	
Trim, cedar	.125	S.F.	.005	.18	.17	.35	
Stain, sealer & 1 coat	1.000	S.F.	.017	.09	.54	.63	
TOTAL			.064	1.36	2.19	3.55	
FIRE RETARDANT, WOOD SHAKES, 8-1/2″ EXPOSURE							
Hand-split red cedar shakes, 18″ long, 8-1/2″ exp.	1.000	S.F.	.038	2.22	1.33	3.55	
#15 asphalt felt paper	1.100	S.F.	.002	.02	.09	.11	
Trim, cedar	.125	S.F.	.005	.18	.17	.35	
Stain, sealer & 1 coat	1.000	S.F.	.017	.09	.54	.63	
TOTAL			.062	2.51	2.13	4.64	

The costs in this system are based on a square foot of wall area.

Shingle Siding System Components

Component Description	QUAN.	UNIT	LABOR-HOURS	COST PER S.F. MAT.	COST PER S.F. INST.	COST PER S.F. TOTAL
Shingles						
Shingles wood, white cedar 16″ long, "A" grade, 5″ exposure	1.000	S.F.	.033	1.23	1.16	2.39
7″ exposure	1.000	S.F.	.030	1.11	1.04	2.15
8-1/2″ exposure	1.000	S.F.	.032	.70	1.11	1.81
10″ exposure	1.000	S.F.	.028	.62	.97	1.59
"B" grade, 5″ exposure	1.000	S.F.	.040	1.21	1.39	2.60
7″ exposure	1.000	S.F.	.028	.85	.97	1.82
8-1/2″ exposure	1.000	S.F.	.024	.73	.83	1.56
10″ exposure	1.000	S.F.	.020	.61	.70	1.31
Fire retardant, "A" grade, 5″ exposure	1.000	S.F.	.033	1.56	1.16	2.72
7″ exposure	1.000	S.F.	.030	1.41	1.04	2.45
8-1/2″ exposure	1.000	S.F.	.027	1.24	.93	2.17
10″ exposure	1.000	S.F.	.023	1.09	.81	1.90
"B" grade, 5″ exposure	1.000	S.F.	.040	1.54	1.39	2.93
7″ exposure	1.000	S.F.	.028	1.08	.97	2.05
8-1/2″ exposure	1.000	S.F.	.024	.93	.83	1.76
10″ exposure	1.000	S.F.	.020	.78	.70	1.48
Hand-split, red cedar, 24″ long, 7″ exposure	1.000	S.F.	.045	2.13	1.55	3.68
8-1/2″ exposure	1.000	S.F.	.038	1.82	1.33	3.15
10″ exposure	1.000	S.F.	.032	1.52	1.11	2.63
12″ exposure	1.000	S.F.	.026	1.22	.89	2.11
Fire retardant, 7″ exposure	1.000	S.F.	.045	2.59	1.55	4.14
18″ long, 5″ exposure	1.000	S.F.	.068	1.82	2.36	4.18
7″ exposure	1.000	S.F.	.048	1.28	1.67	2.95
10″ exposure	1.000	S.F.	.036	.96	1.25	2.21

Metal & Plastic Siding Systems

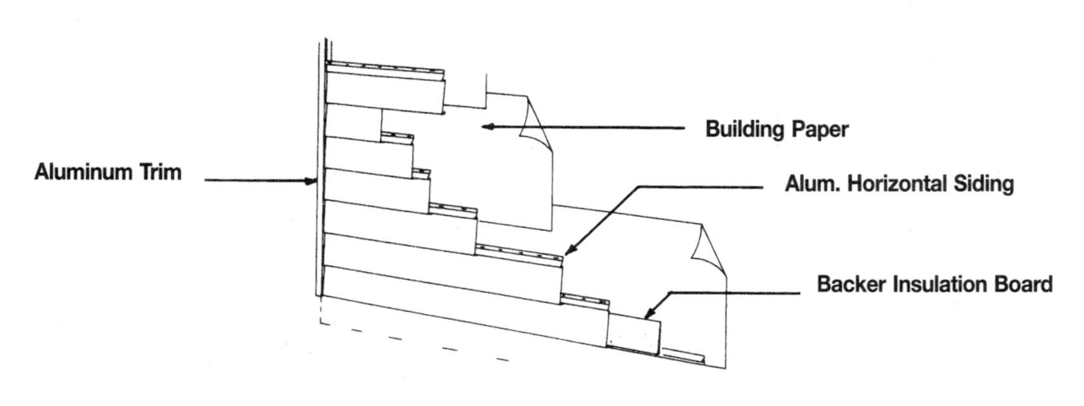

System Description	QUAN.	UNIT	LABOR-HOURS	COST PER S.F. MAT.	COST PER S.F. INST.	COST PER S.F. TOTAL	____S.F. EXTENSIONS
ALUMINUM CLAPBOARD SIDING, 8″ WIDE, WHITE							
Aluminum horizontal siding, 8″ clapboard	1.000	S.F.	.031	1.38	1.08	2.46	
Backer, insulation board	1.000	S.F.	.008	.44	.28	.72	
Trim, aluminum	.600	L.F.	.016	.62	.55	1.17	
Paper, #15 asphalt felt	1.100	S.F.	.002	.02	.09	.11	
TOTAL			.057	2.46	2.00	4.46	
ALUMINUM CLAPBOARD SIDING, DOUBLE 5″ PATTERN							
Aluminum horizontal siding, 10″ clapboard, white	1.000	S.F.	.029	1.31	1.01	2.32	
Backer, insulation board	1.000	S.F.	.008	.44	.28	.72	
Trim, Aluminum	.600	L.F.	.016	.62	.55	1.17	
Paper, #15 asphalt felt	1.100	S.F.	.002	.02	.09	.11	
TOTAL			.055	2.39	1.93	4.32	
ALUMINUM CLAPBOARD SIDING, SHAKE FINISH							
Aluminum horizontal siding, 10″ shake finish, white	1.000	S.F.	.029	1.63	1.01	2.64	
Backer, insulation board	1.000	S.F.	.008	.44	.28	.72	
Trim, Aluminum	.600	L.F.	.016	.62	.55	1.17	
Paper, #15 asphalt felt	1.100	S.F.	.002	.02	.09	.11	
TOTAL			.055	2.71	1.93	4.64	
ALUMINUM VERTICAL BOARD & BATTEN, WHITE							
Aluminum vertical board & batten	1.000	S.F.	.027	1.54	.94	2.48	
Backer insulation board	1.000	S.F.	.008	.44	.28	.72	
Trim, aluminum	.600	L.F.	.016	.62	.55	1.17	
Paper, #15 asphalt felt	1.100	S.F.	.002	.02	.09	.11	
TOTAL			.053	2.62	1.86	4.48	
VINYL CLAPBOARD SIDING, 8″ WIDE, WHITE							
PVC vinyl horizontal siding, 8″ clapboard	1.000	S.F.	.032	.65	1.12	1.77	
Backer, insulation board	1.000	S.F.	.008	.44	.28	.72	
Trim, vinyl	.600	L.F.	.014	.39	.49	.88	
Paper, #15 asphalt felt	1.100	S.F.	.002	.02	.09	.11	
TOTAL			.056	1.50	1.98	3.48	

Metal & Plastic Siding Systems

System Description	QUAN.	UNIT	LABOR-HOURS	COST PER S.F. MAT.	INST.	TOTAL	S.F. EXTENSIONS
VINYL CLAPBOARD SIDING, DOUBLE 5″ PATTERN							
PVC vinyl horizontal siding, 10″ clapboard, white	1.000	S.F.	.029	.57	1.01	1.58	
Backer, insulation board	1.000	S.F.	.008	.44	.28	.72	
Trim, vinyl	.600	L.F.	.014	.39	.49	.88	
Paper, #15 asphalt felt	1.100	S.F.	.002	.02	.09	.11	
TOTAL			.053	1.42	1.87	3.29	
VINYL CLAPBOARD SIDING, EMBOSSED, SINGLE							
PVC vinyl horizontal siding, embossed 8″, white	1.000	S.F.	.032	.73	1.12	1.85	
Backer, insulation board	1.000	S.F.	.008	.44	.28	.72	
Trim, vinyl	.600	L.F.	.014	.39	.49	.88	
Paper, #15 asphalt felt	1.100	S.F.	.002	.02	.09	.11	
TOTAL			.056	1.58	1.98	3.56	
VINYL VERTICAL BOARD & BATTEN, WHITE							
PVC vinyl vertical board & batten	1.000	S.F.	.029	1.51	1.01	2.52	
Backer, insulation board	1.000	S.F.	.008	.44	.28	.72	
Trim, vinyl	.600	L.F.	.014	.39	.49	.88	
Paper, #15 asphalt felt	1.100	S.F.	.002	.02	.09	.11	
TOTAL			.053	2.36	1.87	4.23	

The costs in this system are on a square foot of wall basis.

Metal & Plastic Siding System Components

Component Description	QUAN.	UNIT	LABOR-HOURS	COST PER S.F. MAT.	INST.	TOTAL
Siding						
Siding, aluminum, .024″ thick, smooth, 8″ wide, white	1.000	S.F.	.031	1.38	1.08	2.46
Color	1.000	S.F.	.031	1.47	1.08	2.55
Double 4″ pattern, 8″ wide, white	1.000	S.F.	.031	1.31	1.08	2.39
Color	1.000	S.F.	.031	1.40	1.08	2.48
Embossed, single, 8″ wide, white	1.000	S.F.	.031	1.63	1.08	2.71
Color	1.000	S.F.	.031	1.72	1.08	2.80
Alum siding with insulation board, smooth, 8″ wide, white	1.000	S.F.	.031	1.32	1.08	2.40
Color	1.000	S.F.	.031	1.41	1.08	2.49
Vinyl siding, 8″ wide, smooth, white	1.000	S.F.	.032	.65	1.12	1.77
Color	1.000	S.F.	.032	.74	1.12	1.86
10″ wide, Dutch lap, smooth, white	1.000	S.F.	.029	.70	1.01	1.71
Color	1.000	S.F.	.029	.79	1.01	1.80
Vinyl, shake finish, 10″ wide, white	1.000	S.F.	.029	2.09	1.01	3.10
Color	1.000	S.F.	.029	2.18	1.01	3.19
Backer board						
Backer board, installed in siding panels 8″ or 10″ wide	1.000	S.F.	.008	.44	.28	.72
4′ x 8′ sheets, polystyrene, 3/4″ thick	1.000	S.F.	.010	.35	.35	.70
4′ x 8′ fiberboard, plain	1.000	S.F.	.008	.44	.28	.72
Paper						
Kraft paper, plain	1.100	S.F.	.002	.04	.09	.13
Foil backed	1.100	S.F.	.002	.07	.09	.16

Insulation Systems

Component Description	QUAN.	UNIT	LABOR-HOURS	EXTERIOR WALLS		
				MAT.	INST.	TOTAL
Poured Insulation						
Poured insulation, cellulose fiber, R3.8 per inch (1″ thick)	1.000	S.F.	.003	.04	.12	.16
Fiberglass , R4.0 per inch (1″ thick)	1.000	S.F.	.003	.03	.12	.15
Mineral wool, R3.0 per inch (1″ thick)	1.000	S.F.	.003	.03	.12	.15
Polystyrene, R4.0 per inch (1″ thick)	1.000	S.F.	.003	.19	.12	.31
Vermiculite, R2.7 per inch (1″ thick)	1.000	S.F.	.003	.15	.12	.27
Perlite, R2.7 per inch (1″ thick)	1.000	S.F.	.003	.15	.12	.27
Reflective Insulation						
Reflective insulation, aluminum foil reinforced with scrim	1.000	S.F.	.004	.15	.15	.30
Reinforced with woven polyolefin	1.000	S.F.	.004	.19	.15	.34
With single bubble air space, R8.8	1.000	S.F.	.005	.30	.19	.49
With double bubble air space, R9.8	1.000	S.F.	.005	.32	.19	.51
Rigid Insulation						
Rigid insulation, fiberglass, unfaced,						
1-1/2″ thick, R6.2	1.000	S.F.	.008	.44	.28	.72
2″ thick, R8.3	1.000	S.F.	.008	.47	.28	.75
2-1/2″ thick, R10.3	1.000	S.F.	.010	.59	.35	.94
3″ thick, R12.4	1.000	S.F.	.010	.59	.35	.94
Foil faced, 1″ thick, R4.3	1.000	S.F.	.008	.86	.28	1.14
1-1/2″ thick, R6.2	1.000	S.F.	.008	1.16	.28	1.44
2″ thick, R8.7	1.000	S.F.	.009	1.44	.31	1.75
2-1/2″ thick, R10.9	1.000	S.F.	.010	1.71	.35	2.06
3″ thick, R13.0	1.000	S.F.	.010	1.86	.35	2.21
Foam glass, 1-1/2″ thick R2.64	1.000	S.F.	.010	1.68	.35	2.03
2″ thick R5.26	1.000	S.F.	.011	2.98	.38	3.36
Perlite, 1″ thick R2.77	1.000	S.F.	.010	.28	.35	.63
2″ thick R5.55	1.000	S.F.	.011	.54	.38	.92
Polystyrene, extruded, blue, 2.2#/C.F., 3/4″ thick R4	1.000	S.F.	.010	.35	.35	.70
1-1/2″ thick R8.1	1.000	S.F.	.011	.70	.38	1.08
2″ thick R10.8	1.000	S.F.	.011	1.00	.38	1.38
Molded bead board, white, 1″ thick R3.85	1.000	S.F.	.010	.14	.35	.49
1-1/2″ thick, R5.6	1.000	S.F.	.011	.39	.38	.77
2″ thick, R7.7	1.000	S.F.	.011	.55	.38	.93
Non-rigid insulation, batts						
Fiberglass, kraft faced, 3-1/2″ thick, R11, 11″ wide	1.000	S.F.	.005	.24	.17	.41
15″ wide	1.000	S.F.	.005	.24	.17	.41
23″ wide	1.000	S.F.	.005	.24	.17	.41
6″ thick, R19, 11″ wide	1.000	S.F.	.006	.35	.21	.56
15″ wide	1.000	S.F.	.006	.35	.21	.56
23″ wide	1.000	S.F.	.006	.35	.21	.56
9″ thick, R30, 15″ wide	1.000	S.F.	.006	.64	.21	.85
23″ wide	1.000	S.F.	.006	.64	.21	.85
12″ thick, R38, 15″ wide	1.000	S.F.	.006	.81	.21	1.02
23″ wide	1.000	S.F.	.006	.81	.21	1.02
Fiberglass, foil faced, 3-1/2″ thick, R11, 15″ wide	1.000	S.F.	.005	.36	.17	.53
23″ wide	1.000	S.F.	.005	.36	.17	.53
6″ thick, R19, 15″ thick	1.000	S.F.	.005	.44	.17	.61
23″ wide	1.000	S.F.	.005	.44	.17	.61
9″ thick, R30, 15″ wide	1.000	S.F.	.006	.76	.21	.97
Non-rigid insulation batts						
Fiberglass unfaced, 3-1/2″ thick, R11, 15″ wide	1.000	S.F.	.005	.22	.17	.39

Insulation Systems

Component Description	QUAN.	UNIT	LABOR-HOURS	EXTERIOR WALLS		
				MAT.	INST.	TOTAL
23″ wide	1.000	S.F.	.005	.22	.17	.39
6″ thick, R19, 15″ wide	1.000	S.F.	.006	.36	.21	.57
23″ wide	1.000	S.F.	.006	.36	.21	.57
9″ thick, R19, 15″ wide	1.000	S.F.	.007	.64	.24	.88
23″ wide	1.000	S.F.	.007	.64	.24	.88
12″ thick, R38, 15″ wide	1.000	S.F.	.007	.81	.24	1.05
23″ wide	1.000	S.F.	.007	.81	.24	1.05
Mineral fiber batts, 3″ thick, R11	1.000	S.F.	.005	.28	.17	.45
3-1/2″ thick, R13	1.000	S.F.	.005	.28	.17	.45
6″ thick, R19	1.000	S.F.	.005	.42	.17	.59
6-1/2″ thick, R22	1.000	S.F.	.005	.42	.17	.59
10″ thick, R30	1.000	S.F.	.006	.66	.21	.87

Double Hung Window Systems

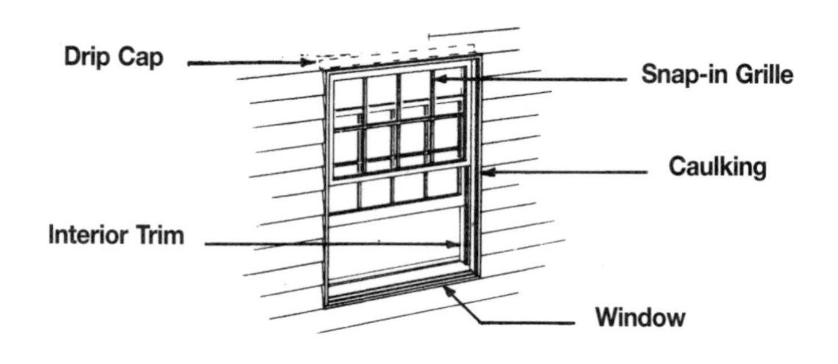

System Description	QUAN.	UNIT	LABOR-HOURS	COST EACH MAT.	COST EACH INST.	COST EACH TOTAL	EA. EXTENSIONS
BUILDER'S QUALITY WOOD WINDOW 2' X 3', DOUBLE HUNG							
Window, primed, builder's quality, 2' x 3', insulating glass	1.000	Ea.	.800	234.00	28.00	262.00	
Trim, interior casing	11.000	L.F.	.367	9.46	12.76	22.22	
Paint, interior & exterior, primer & 2 coats	2.000	Face	1.778	2.02	55.00	57.02	
Caulking	10.000	L.F.	.323	1.60	11.30	12.90	
Snap-in grille	1.000	Set	.333	41.00	11.60	52.60	
Drip cap, metal	2.000	L.F.	.040	.44	1.40	1.84	
TOTAL			3.641	288.52	120.06	408.58	
BUILDER'S QUALITY WOOD WINDOW 3' X 4' DOUBLE HUNG							
Window, builder's quality, 3' x 4', insulating glass	1.000	Ea.	.889	315.00	31.00	346.00	
Trim, interior casing	15.000	L.F.	.367	9.46	12.76	22.22	
Paint, interior & exterior, primer & 2 coats	2.000	Face	1.778	2.02	55.00	57.02	
Caulking	14.000	L.F.	.323	1.60	11.30	12.90	
Snap-in grille	1.000	Set	.333	41.00	11.60	52.60	
Drip cap, metal	3.000	L.F.	.040	.44	1.40	1.84	
TOTAL			3.730	369.52	123.06	492.58	
PLASTIC CLAD WOOD WINDOW 3' X 4', DOUBLE HUNG							
Window, plastic clad, premium, 3' x 4', insulating glass	1.000	Ea.	.889	305.00	31.00	336.00	
Trim, interior casing	15.000	L.F.	.500	12.90	17.40	30.30	
Paint, interior, primer & 2 coats	1.000	Face	.889	1.01	27.50	28.51	
Caulking	14.000	L.F.	.452	2.24	15.82	18.06	
Snap-in grille	1.000	Set	.333	41.00	11.60	52.60	
TOTAL			3.063	362.15	103.32	465.47	
METAL CLAD WOOD WINDOW 3' X 4' DOUBLE HUNG							
Window, deluxe, 3' x 4' insulating glass	1.000	Ea.	.889	242.00	31.00	273.00	
Trim interior casing	15.000	L.F.	.567	14.62	19.72	34.34	
Paint interior, primer & 2 coats	1.000	Face	.889	1.01	27.50	28.51	
Caulking	14.000	L.F.	.516	2.56	18.08	20.64	
Snap-in grille	1.000	Set	.235	115.00	8.20	123.20	
Drip cap, metal	3.000	L.F.	.060	.66	2.10	2.76	
TOTAL			3.156	375.85	106.60	482.45	

Double Hung Window Systems

System Description	QUAN.	UNIT	LABOR-HOURS	COST EACH MAT.	COST EACH INST.	COST EACH TOTAL	EA. EXTENSIONS
PLASTIC CLAD WOOD WINDOW 3'-6" X 6' DOUBLE HUNG							
Window, premium 3'-6" x 6' insulating glass	1.000	Ea.	1.000	410.00	35.00	445.00	
Trim interior casing	20.000	L.F.	.667	17.20	23.20	40.40	
Paint, interior, primer & 2 coats	1.000	Face	.889	1.01	27.50	28.51	
Caulking	19.000	L.F.	.613	3.04	21.47	24.51	
Snap-in grille	1.000	Set	.235	115.00	8.20	123.20	
TOTAL			3.404	546.25	115.37	661.62	
METAL CLAD WOOD WINDOW, 3' X 5', DOUBLE HUNG							
Window, metal clad, deluxe, 3' x 5', insulating glass	1.000	Ea.	1.000	280.00	35.00	315.00	
Trim, interior casing	17.000	L.F.	.567	14.62	19.72	34.34	
Paint, interior, primer & 2 coats	1.000	Face	.889	1.01	27.50	28.51	
Caulking	16.000	L.F.	.516	2.56	18.08	20.64	
Snap-in grille	1.000	Set	.235	115.00	8.20	123.20	
Drip cap, metal	3.000	L.F.	.060	.66	2.10	2.76	
TOTAL			3.267	413.85	110.60	524.45	
BUILDER'S QUALITY WOOD WINDOW 4' X 4'-6" DOUBLE HUNG							
Window, builder's quality, 4' x 4'-6", insulating glass	1.000	Ea.	1.000	355.00	35.00	390.00	
Trim, interior casing	18.000	L.F.	.600	15.48	20.88	36.36	
Paint interior & exterior, primer & 2 coats	2.000	Face	1.778	2.02	55.00	57.02	
Caulking	17.000	L.F.	.548	2.72	19.21	21.93	
Snap-in grille	1.000	Set	.235	115.00	8.20	123.20	
Drip cap, metal	4.000	L.F.	.080	.88	2.80	3.68	
TOTAL			4.241	491.10	141.09	632.19	
PLASTIC CLAD WOOD WINDOW 2'-6" X 3' DOUBLE HUNG							
Window, premium, 2'-6" x 3', insulating glass	1.000	Ea.	.800	234.00	28.00	262.00	
Trim interior casing	12.000	L.F.	.400	10.32	13.92	24.24	
Paint, interior, primer & 2 coats	1.000	Face	.889	1.01	27.50	28.51	
Caulking	11.000	L.F.	.355	1.76	12.43	14.19	
Snap-in grille	1.000	Set	.333	41.00	11.60	52.60	
TOTAL			2.777	288.09	93.45	381.54	

The cost of this system is on a cost per each window basis.

Double Hung Window System Components

Component Description	QUAN.	UNIT	LABOR-HOURS	COST EACH MAT.	COST EACH INST.	COST EACH TOTAL
Windows, double-hung						
Windows, double-hung, builder's quality, 2' x 3', single glass	1.000	Ea.	.800	223.00	28.00	251.00
3' x 4', single glass	1.000	Ea.	.889	305.00	31.00	336.00
Plastic clad premium insulating glass, 2'-6" x 3'	1.000	Ea.	.800	233.00	28.00	261.00
3' x 3'-6"	1.000	Ea.	.800	278.00	28.00	306.00
Metal clad deluxe insulating glass, 2'-6" x 3'	1.000	Ea.	.800	192.00	28.00	220.00
3' x 3'-6"	1.000	Ea.	.800	228.00	28.00	256.00
Trim, interior casing						
Trim, interior casing, window 2' x 3'	11.000	L.F.	.367	9.45	12.75	22.20
3' x 3'-6"	14.000	L.F.	.467	12.05	16.25	28.30

Casement Window Systems

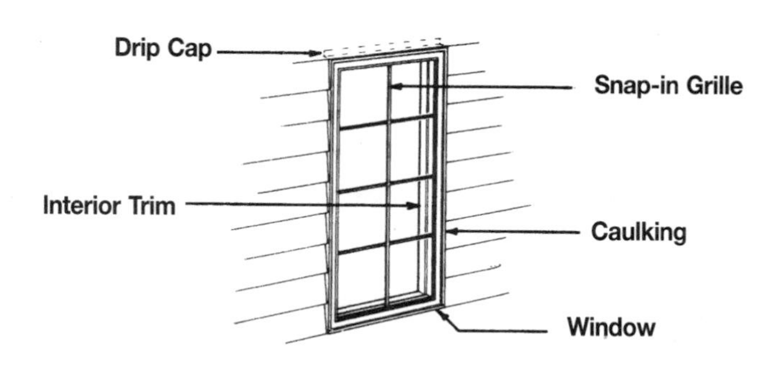

System Description	QUAN.	UNIT	LABOR-HOURS	COST EACH MAT.	COST EACH INST.	COST EACH TOTAL	____EA. EXTENSIONS
BUILDER'S QUALITY WINDOW, WOOD, 2′ BY 3′, CASEMENT							
Window, primed, builder's quality, 2′ x 3′, insulating glass	1.000	Ea.	.800	224.00	28.00	252.00	
Trim, interior casing	11.000	L.F.	.367	9.46	12.76	22.22	
Paint, interior & exterior, primer & 2 coats	2.000	Face	1.778	2.02	55.00	57.02	
Caulking	10.000	L.F.	.323	1.60	11.30	12.90	
Snap-in grille	1.000	Ea.	.267	23.00	9.30	32.30	
Drip cap, metal	2.000	L.F.	.040	.44	1.40	1.84	
TOTAL			3.575	260.52	117.76	378.28	
PLASTIC CLAD WOOD WINDOW 2′ X 3′ CASEMENT							
Window, premium, 2′ x 3′, insulating glass	1.000	Ea.	.800	233.00	28.00	261.00	
Trim, interior casing	11.000	L.F.	.367	9.46	12.76	22.22	
Paint, interior, primer & 2 coats	1.000	Face	.889	1.01	27.50	28.51	
Caulking	10.000	L.F.	.323	1.60	11.30	12.90	
Snap-in grille	1.000	Set	.267	23.00	9.30	32.30	
TOTAL			2.646	268.07	88.86	356.93	
METAL CLAD WOOD WINDOW 2′ X 3′ CASEMENT							
Window, deluxe, 2′ x 3′ insulating glass	1.000	Ea.	.800	182.00	28.00	210.00	
Trim, interior casing	11.000	L.F.	.367	9.46	12.76	22.22	
Paint, interior, primer & 2 coats	1.000	Face	.889	1.01	27.50	28.51	
Caulking	10.000	L.F.	.323	1.60	11.30	12.90	
Snap-in grille	1.000	Set	.267	23.00	9.30	32.30	
Drip cap, metal	2.000	L.F.	.040	.44	1.40	1.84	
TOTAL			2.686	217.51	90.26	307.77	
BUILDER'S QUALITY WOOD WINDOW 2′ X 4′-6″ CASEMENT							
Window, builder's quality, 2′ x 4′-6″, insulating glass	1.000	Ea.	.889	232.00	31.00	263.00	
Trim, interior casing	14.000	L.F.	.467	12.04	16.24	28.28	
Paint interior & exterior, primer & 2 coats	2.000	Face	1.778	2.02	55.00	57.02	
Caulking	13.000	L.F.	.419	2.08	14.69	16.77	
Snap-in grille	1.000	Set	.250	33.50	8.70	42.20	
Drip cap, metal	2.000	L.F.	.040	.44	1.40	1.84	
TOTAL			3.843	282.08	127.03	409.11	

Casement Window Systems

System Description	QUAN.	UNIT	LABOR-HOURS	COST EACH MAT.	COST EACH INST.	COST EACH TOTAL	____EA. EXTENSIONS
PLASTIC CLAD WOOD WINDOW, 2′ X 4′, CASEMENT							
Window, plastic clad, premium, 2′ x 4′, insulating glass	1.000	Ea.	.889	276.00	31.00	307.00	
Trim, interior casing	13.000	L.F.	.433	11.18	15.08	26.26	
Paint, interior, primer & 2 coats	1.000	Ea.	.889	1.01	27.50	28.51	
Caulking	12.000	L.F.	.387	1.92	13.56	15.48	
Snap-in grille	1.000	Ea.	.267	23.00	9.30	32.30	
TOTAL			2.865	313.11	96.44	409.55	
METAL CLAD WOOD WINDOW 2′ X 4′ CASEMENT							
Window, deluxe, 2′ x 4′, insulating glass	1.000	Ea.	.889	219.00	31.00	250.00	
Trim, interior casing	13.000	L.F.	.433	11.18	15.08	26.26	
Paint, interior, primer & 2 coats	1.000	Face	.889	1.01	27.50	28.51	
Caulking	12.000	L.F.	.387	1.92	13.56	15.48	
Snap-in grille	1.000	Set	.267	23.00	9.30	32.30	
Drip cap, metal	2.000	L.F.	.040	.44	1.40	1.84	
TOTAL			2.905	256.55	97.84	354.39	
BUILDER'S QUALITY WOOD WINDOW 2′ X 6′ CASEMENT							
Window, builder's quality, 2′ x 6′ insulating glass	1.000	Ea.	1.000	259.00	35.00	294.00	
Trim, interior casing	17.000	L.F.	.567	14.62	19.72	34.34	
Paint, interior & exterior, primer & 2 coats	2.000	Face	1.778	2.02	55.00	57.02	
Caulking	16.000	L.F.	.516	2.56	18.08	20.64	
Snap-in grille	1.000	Set	.250	33.50	8.70	42.20	
Drip cap, metal	2.000	L.F.	.040	.44	1.40	1.84	
TOTAL			4.151	312.14	137.90	450.04	
METAL CLAD WOOD WINDOW 2′ X 6′ CASEMENT							
Window, deluxe, 2′ x 6′, insulating glass	1.000	Ea.	1.000	285.00	35.00	320.00	
Trim, interior casing	17.000	L.F.	.567	14.62	19.72	34.34	
Paint, interior, primer & 2 coats	1.000	Face	.889	1.01	27.50	28.51	
Caulking	16.000	L.F.	.516	2.56	18.08	20.64	
Snap-in grille	1.000	Set	.250	33.50	8.70	42.20	
Drip cap, metal	2.000	L.F.	.040	.44	1.40	1.84	
TOTAL			3.262	337.13	110.40	447.53	

The cost of this system is on a cost per each window basis.

Casement Window System Components

Component Description	QUAN.	UNIT	LABOR-HOURS	COST EACH MAT.	COST EACH INST.	COST EACH TOTAL
Window, casement						
Window, casement, builders quality, 2′ x 3′, single glass	1.000	Ea.	.800	178.00	28.00	206.00
2′ x 4′-6″, single glass	1.000	Ea.	.727	680.00	25.50	705.50
Plastic clad premium insulating glass, 2′ x 3′	1.000	Ea.	.800	182.00	28.00	210.00
2′ x 5′	1.000	Ea.	1.000	263.00	35.00	298.00
Paint or stain						
Paint or stain, interior or exterior, 2′ x 3′ window, 1 coat	1.000	Face	.444	.37	13.70	14.07
2 coats	1.000	Face	.727	.74	22.50	23.24

Awning Window Systems

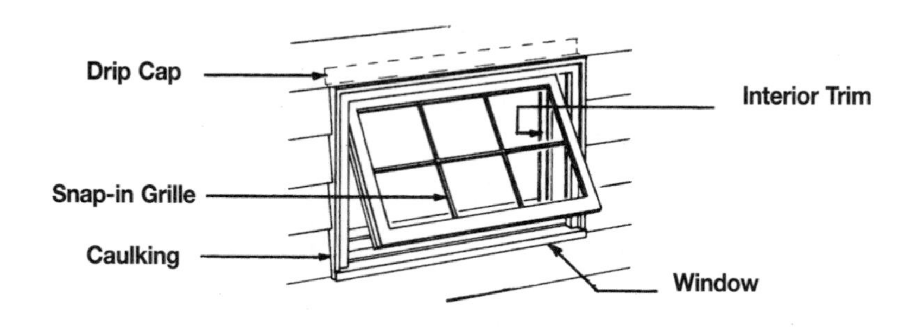

System Description	QUAN.	UNIT	LABOR-HOURS	COST EACH MAT.	COST EACH INST.	COST EACH TOTAL	____EA. EXTENSIONS
BUILDER'S QUALITY WINDOW, WOOD, 34" X 22", AWNING							
Window, builder quality, 34" x 22", insulating glass	1.000	Ea.	.800	228.00	28.00	256.00	
Trim, interior casing	10.500	L.F.	.350	9.03	12.18	21.21	
Paint, interior & exterior, primer & 2 coats	2.000	Face	1.778	2.02	55.00	57.02	
Caulking	9.500	L.F.	.306	1.52	10.74	12.26	
Snap-in grille	1.000	Ea.	.267	18.90	9.30	28.20	
Drip cap, metal	3.000	L.F.	.060	.66	2.10	2.76	
TOTAL			3.561	260.13	117.32	377.45	
PLASTIC CLAD WOOD WINDOW 34" X 22" AWNING							
Window, premium 34" x 22", insulating glass	1.000	Ea.	.800	227.00	28.00	255.00	
Trim, interior casing	10.500	L.F.	.350	9.03	12.18	21.21	
Paint, interior, primer & 2 coats	1.000	Face	.889	1.01	27.50	28.51	
Caulking	9.500	L.F.	.306	1.52	10.74	12.26	
Snap-in grille	1.000	Set	.267	18.90	9.30	28.20	
TOTAL			2.612	257.46	87.72	345.18	
METAL CLAD WOOD WINDOW 34" X 22" AWNING							
Window, deluxe, 34" x 22", insulating glass	1.000	Ea.	.800	203.00	28.00	231.00	
Trim, interior casing	10.500	L.F.	.350	9.03	12.18	21.21	
Paint, interior, primer & 2 coats	1.000	Face	.889	1.01	27.50	28.51	
Caulking	9.500	L.F.	.306	1.52	10.74	12.26	
Snap-in grille	1.000	Ea.	.267	18.90	9.30	28.20	
Drip cap, metal	3.000		.060	.66	2.10	2.76	
TOTAL			2.672	234.12	89.82	323.94	
BUILDER'S QUALITY WINDOW, WOOD, 40" X 28", AWNING							
Window, builder's quality, 40" x 28", insulating glass	1.000	Ea.	.889	290.00	31.00	321.00	
Trim, interior casing	13.500	L.F.	.450	11.61	15.66	27.27	
Paint interior & exterior primer & 2 coats	2.000	Face	1.778	2.02	55.00	57.02	
Caulking	12.500	L.F.	.403	2.00	14.13	16.13	
Snap-in grille	1.000	Set	.250	27.50	8.70	36.20	
Drip cap, metal	3.500	L.F.	.080	.88	2.80	3.68	
TOTAL			3.850	334.01	127.29	461.30	

Awning Window Systems

System Description	QUAN.	UNIT	LABOR-HOURS	COST EACH			___EA. EXTENSIONS
				MAT.	INST.	TOTAL	
PLASTIC CLAD WOOD WINDOW, 40" X 28", AWNING							
Window, plastic clad, premium, 40" x 28", insulating glass	1.000	Ea.	.889	290.00	31.00	321.00	
Trim interior casing	13.500	L.F.	.450	11.61	15.66	27.27	
Paint, interior, primer & 2 coats	1.000	Face	.889	1.01	27.50	28.51	
Caulking	12.500	L.F.	.403	2.00	14.13	16.13	
Snap-in grille	1.000	Ea.	.267	18.90	9.30	28.20	
TOTAL			2.898	323.52	97.59	421.11	
METAL CLAD WOOD WINDOW 40" X 28" AWNING							
Window, deluxe, 40" x 28", insulating glass	1.000	Ea.	.889	298.00	31.00	329.00	
Trim, interior casing	13.500	L.F.	.450	11.61	15.66	27.27	
Paint, interior, primer & 2 coats	1.000	Face	.889	1.01	27.50	28.51	
Caulking	12.500	L.F.	.403	2.00	14.13	16.13	
Snap-in grille	1.000	Set	.267	18.90	9.30	28.20	
Drip cap, metal	3.500	L.F.	.080	.88	2.80	3.68	
TOTAL			2.978	332.40	100.39	432.79	
PLASTIC CLAD WOOD WINDOW 48" X 36" AWNING							
Window, premium, 48" x 36" insulating glass	1.000	Ea.	1.000	315.00	35.00	350.00	
Trim, interior casing	15.000	L.F.	.500	12.90	17.40	30.30	
Paint, interior, primer & 2 coats	1.000	Face	.889	1.01	27.50	28.51	
Caulking	14.000	L.F.	.452	2.24	15.82	18.06	
Snap-in grille	1.000	Ea.	.250	27.50	8.70	36.20	
TOTAL			3.091	358.65	104.42	463.07	
METAL CLAD WOOD WINDOW, 48" X 36", AWNING							
Window, metal clad, deluxe, 48" x 36", insulating glass	1.000	Ea.	1.000	325.00	35.00	360.00	
Trim, interior casing	15.000	L.F.	.500	12.90	17.40	30.30	
Paint, interior, primer & 2 coats	1.000	Face	.889	1.01	27.50	28.51	
Caulking	14.000	L.F.	.452	2.24	15.82	18.06	
Snap-in grille	1.000	Ea.	.250	27.50	8.70	36.20	
Drip cap, metal	4.000	L.F.	.080	.88	2.80	3.68	
TOTAL			3.171	369.53	107.22	476.75	

The cost of this system is on a cost per each window basis.

Awning Window System Components

Component Description	QUAN.	UNIT	LABOR-HOURS	COST EACH		
				MAT.	INST.	TOTAL
Windows, awning						
Windows, awning, builder's quality, 34" x 22", insulated glass	1.000	Ea.	.800	216.00	28.00	244.00
40" x 28", insulated glass	1.000	Ea.	.889	275.00	31.00	306.00
Plastic clad premium insulating glass, 34" x 22"	1.000	Ea.	.800	227.00	28.00	255.00
40" x 22"	1.000	Ea.	.800	250.00	28.00	278.00
36" x 36"	1.000	Ea.	.889	290.00	31.00	321.00
48" x 28"	1.000	Ea.	1.000	315.00	35.00	350.00
Metal clad deluxe insulating glass, 34" x 22"	1.000	Ea.	.800	203.00	28.00	231.00
40" x 22"	1.000	Ea.	.800	251.00	28.00	279.00
36" x 25"	1.000	Ea.	.889	239.00	31.00	270.00
40" x 30"	1.000	Ea.	.889	298.00	31.00	329.00

Sliding Window Systems

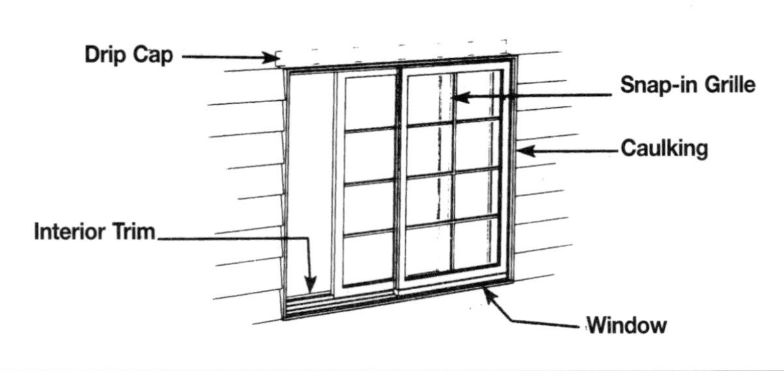

System Description	QUAN.	UNIT	LABOR-HOURS	COST EACH MAT.	COST EACH INST.	COST EACH TOTAL	___EA. EXTENSIONS
BUILDER'S QUALITY WOOD WINDOW, 3' X 2', SLIDING							
Window, primed, builder's quality, 3' x 2', insul. glass	1.000	Ea.	.800	182.00	28.00	210.00	
Trim, interior casing	11.000	L.F.	.367	9.46	12.76	22.22	
Paint, interior & exterior, primer & 2 coats	2.000	Face	1.778	2.02	55.00	57.02	
Caulking	10.000	L.F.	.323	1.60	11.30	12.90	
Snap-in grille	1.000	Set	.333	22.50	11.60	34.10	
Drip cap, metal	3.000	L.F.	.060	.66	2.10	2.76	
TOTAL			3.661	218.24	120.76	339.00	
PLASTIC CLAD WOOD WINDOW 3' X 3' SLIDING							
Window, premium, 3' x 3', insulating glass	1.000	Ea.	.800	525.00	28.00	553.00	
Trim, interior casing	13.000	L.F.	.433	11.18	15.08	26.26	
Paint, interior, primer & 2 coats	1.000	Face	.889	1.01	27.50	28.51	
Caulking	12.000	L.F.	.387	1.92	13.56	15.48	
Snap-in grille	1.000	Set	.333	22.50	11.60	34.10	
TOTAL			2.842	561.61	95.74	657.35	
METAL CLAD WOOD WINDOW 3' X 3' SLIDING							
Window, deluxe, 3' x 3', insulating glass	1.000	Ea.	.800	293.00	28.00	321.00	
Trim, interior casing	13.000	L.F.	.433	11.18	15.08	26.26	
Paint, interior, primer & 2 coats	1.000	Face	.889	1.01	27.50	28.51	
Caulking	12.000	L.F.	.387	1.92	13.56	15.48	
Snap-in grille	1.000	Set	.333	22.50	11.60	34.10	
Drip cap, metal	3.000	L.F.	.060	.66	2.10	2.76	
TOTAL			2.902	330.27	97.84	428.11	
BUILDER'S QUALITY WOOD WINDOW 4' X 3'-6" SLIDING							
Window, builder's quality, 4' x 3'-6", insulating glass	1.000	Ea.	.889	215.00	31.00	246.00	
Trim, interior casing	16.000	L.F.	.533	13.76	18.56	32.32	
Paint, interior & exterior, primer & 2 coats	2.000	Face	1.778	2.02	55.00	57.02	
Caulking	15.000	L.F.	.548	2.72	19.21	21.93	
Snap-in grille	1.000	Set	.333	22.50	11.60	34.10	
Drip cap, metal	4.000	L.F.	.080	.88	2.80	3.68	
TOTAL			4.161	256.88	138.17	395.05	

Sliding Window Systems

System Description	QUAN.	UNIT	LABOR-HOURS	COST EACH MAT.	COST EACH INST.	COST EACH TOTAL	____EA. EXTENSIONS
PLASTIC CLAD WOOD WINDOW, 4′ X 3′-6″, SLIDING							
Window, plastic clad, premium, 4′ x 3′-6″, insulating glass	1.000	Ea.	.889	630.00	31.00	661.00	
Trim, interior casing	16.000	L.F.	.533	13.76	18.56	32.32	
Paint, interior, primer & 2 coats	1.000	Face	.889	1.01	27.50	28.51	
Caulking	17.000	L.F.	.548	2.72	19.21	21.93	
Snap-in grille	1.000	Set	.333	22.50	11.60	34.10	
TOTAL			3.192	669.99	107.87	777.86	
METAL CLAD WOOD WINDOW 4′ X 3′-6″ SLIDING							
Window, deluxe, 4′ x 3′-6″, insulating glass	1.000	Ea.	.889	360.00	31.00	391.00	
Trim, interior casing	16.000	L.F.	.533	13.76	18.56	32.32	
Paint, interior, primer & 2 coats	1.000	Face	.889	1.01	27.50	28.51	
Caulking	15.000	L.F.	.548	2.72	19.21	21.93	
Snap-in grille	1.000	Set	.333	22.50	11.60	34.10	
Drip cap, metal	4.000	L.F.	.080	.88	2.80	3.68	
TOTAL			3.272	400.87	110.67	511.54	
BUILDER'S QUALITY WOOD WINDOW 6′ X 5′ SLIDING							
Window, builder's quality, 6′ x 5′, insulating glass	1.000	Ea.	1.000	375.00	35.00	410.00	
Trim, interior casing	23.000	L.F.	.767	19.78	26.68	46.46	
Paint, interior & exterior, primer & 2 coats	2.000	Face	1.778	2.02	55.00	57.02	
Caulking	22.000	L.F.	.710	3.52	24.86	28.38	
Snap-in grille	1.000	Set	.364	46.50	12.65	59.15	
Drip cap, metal	6.000	L.F.	.120	1.32	4.20	5.52	
TOTAL			4.739	448.14	158.39	606.53	
METAL CLAD WOOD WINDOW, 6′ X 5′, SLIDING							
Window, metal clad, deluxe, 6′ x 5′, insulating glass	1.000	Ea.	1.000	530.00	35.00	565.00	
Trim, interior casing	23.000	L.F.	.767	19.78	26.68	46.46	
Paint, interior, primer & 2 coats	1.000	Face	.889	1.01	27.50	28.51	
Caulking	22.000	L.F.	.710	3.52	24.86	28.38	
Snap-in grille	1.000	Set	.364	46.50	12.65	59.15	
Drip cap, metal	6.000	L.F.	.120	1.32	4.20	5.52	
TOTAL			3.850	602.13	130.89	733.02	

The cost of this system is on a cost per each window basis.

Sliding Window System Components

Component Description	QUAN.	UNIT	LABOR-HOURS	COST EACH MAT.	COST EACH INST.	COST EACH TOTAL
Windows, sliding						
Windows, sliding, builder's quality, 3′ x 3′, single glass	1.000	Ea.	.800	144.00	28.00	172.00
6′ x 5′, single glass	1.000	Ea.	1.000	315.00	35.00	350.00
Plastic clad premium insulating glass, 3′ x 3′	1.000	Ea.	.800	525.00	28.00	553.00
5′ x 4′	1.000	Ea.	.889	730.00	31.00	761.00
6′ x 5′	1.000	Ea.	1.000	910.00	35.00	945.00
Metal clad deluxe insulating glass, 3′ x 3′	1.000	Ea.	.800	293.00	28.00	321.00
5′ x 4′	1.000	Ea.	.889	435.00	31.00	466.00

Bow/Bay Window Systems

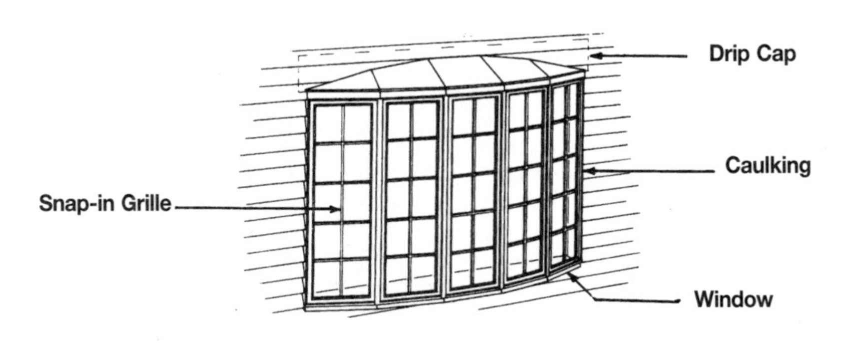

System Description	QUAN.	UNIT	LABOR-HOURS	COST EACH MAT.	COST EACH INST.	COST EACH TOTAL	___EA. EXTENSIONS
AWNING TYPE BOW WINDOW, BUILDER'S QUALITY, 8' X 5'							
Window, primed, builder's quality, 8' x 5', insulating glass	1.000	Ea.	1.600	1225.00	55.50	1280.50	
Trim, interior casing	27.000	L.F.	.900	23.22	31.32	54.54	
Paint, interior & exterior, primer & 1 coat	2.000	Face	3.200	10.80	99.00	109.80	
Drip cap, vinyl	1.000	Ea.	.533	75.50	18.55	94.05	
Caulking	26.000	L.F.	.839	4.16	29.38	33.54	
Snap-in grilles	1.000	Set	1.067	92.00	37.20	129.20	
TOTAL			8.139	1430.68	270.95	1701.63	
AWNING TYPE BOW WINDOW PLASTIC CLAD 10' X 5'							
Window, premium, 10' x 5', insulating glass	1.000	Ea.	2.286	2350.00	79.50	2429.50	
Trim, interior casing	33.000	L.F.	1.100	28.38	38.28	66.66	
Paint, interior, primer & 2 coats	1.000	Face	1.778	2.02	55.00	57.02	
Drip cap, vinyl	1.000	Ea.	.615	82.50	21.50	104.00	
Caulking	32.000	L.F.	1.032	5.12	36.16	41.28	
Snap-in grille	5.000	Set	1.333	115.00	46.50	161.50	
TOTAL			8.144	2583.02	276.94	2859.96	
CASEMENT TYPE BOW WINDOW BUILDER'S QUALITY 12' X 6'							
Window, builder's quality, 12' x 6', insulating glass	1.000	Ea.	2.667	1900.00	93.00	1993.00	
Trim, interior casing	37.000	L.F.	1.233	31.82	42.92	74.74	
Paint, interior & exterior, primer & 2 coats	2.000	Face	2.667	7.24	82.00	89.24	
Drip cap, vinyl	1.000	Ea.	.615	82.50	21.50	104.00	
Caulking	36.000	L.F.	1.161	5.76	40.68	46.44	
Snap-in grille	6.000	Set	1.600	138.00	55.80	193.80	
TOTAL			9.943	2165.32	335.90	2501.22	
CASEMENT TYPE BOW WINDOW, PLASTIC CLAD, 10' X 6'							
Window, plastic clad, premium, 10' x 6', insulating glass	1.000	Ea.	2.286	1875.00	79.50	1954.50	
Trim, interior casing	33.000	L.F.	1.100	28.38	38.28	66.66	
Paint, interior, primer & 1 coat	1.000	Face	1.778	2.02	55.00	57.02	
Drip cap, vinyl	1.000	Ea.	.615	82.50	21.50	104.00	
Caulking	32.000	L.F.	1.032	5.12	36.16	41.28	
Snap-in grilles	1.000	Set	1.333	115.00	46.50	161.50	
TOTAL			8.144	2108.02	276.94	2384.96	

Bow/Bay Window Systems

System Description	QUAN.	UNIT	LABOR-HOURS	COST EACH MAT.	COST EACH INST.	COST EACH TOTAL	____EA. EXTENSIONS
CASEMENT TYPE BOW WINDOW METAL CLAD 12′ X 6′							
Window, deluxe, 12′ x 6′, insulating glass	1.000	Ea.	2.667	2300.00	93.00	2393.00	
Trim, interior casing	37.000	L.F.	1.233	31.82	42.92	74.74	
Paint, interior, primer & 2 coats	1.000	Face	2.667	7.24	82.00	89.24	
Drip cap, vinyl	1.000	Ea.	.615	82.50	21.50	104.00	
Caulking	36.000	L.F.	1.161	5.76	40.68	46.44	
Snap-in grille	6.000	Set	1.600	138.00	55.80	193.80	
TOTAL			9.943	2565.32	335.90	2901.22	
DOUBLE HUNG TYPE BOW WINDOW BUILDER'S QUALITY 9′ X 5′							
Window, builder's quality 9′ x 5′ insulating glass	1.000	Ea.	1.600	1100.00	55.50	1155.50	
Trim interior casing	27.000	L.F.	.967	24.94	33.64	58.58	
Paint interior & exterior primer & 2 coats	2.000	Face	3.556	4.04	110.00	114.04	
Drip cap vinyl	1.000	Ea.	.615	82.50	21.50	104.00	
Caulking	26.000	L.F.	.903	4.48	31.64	36.12	
Snap-in grille	4.000	Set	1.067	92.00	37.20	129.20	
TOTAL			8.708	1307.96	289.48	1597.44	
DOUBLE HUNG TYPE BOW WINDOW PLASTIC CLAD 9′ X 5′							
Window, premium 9′ x 5′ insulating glass	1.000	Ea.	2.667	1175.00	93.00	1268.00	
Trim interior casing	27.000	L.F.	.967	24.94	33.64	58.58	
Paint interior primer & 2 coats	1.000	Face	1.778	2.02	55.00	57.02	
Drip cap vinyl	1.000	Ea.	.615	82.50	21.50	104.00	
Caulking	26.000	L.F.	.903	4.48	31.64	36.12	
Snap-in grille	4.000	Set	1.067	92.00	37.20	129.20	
TOTAL			7.997	1380.94	271.98	1652.92	
DOUBLE HUNG TYPE, METAL CLAD, 9′ X 5′							
Window, metal clad, deluxe, 9′ x 5′, insulating glass	1.000	Ea.	2.667	1125.00	93.00	1218.00	
Trim, interior casing	29.000	L.F.	.967	24.94	33.64	58.58	
Paint, interior, primer & 1 coat	1.000	Face	1.778	2.02	55.00	57.02	
Drip cap, vinyl	1.000	Set	.615	82.50	21.50	104.00	
Caulking	28.000	L.F.	.903	4.48	31.64	36.12	
Snap-in grilles	1.000	Set	1.067	92.00	37.20	129.20	
TOTAL			7.997	1330.94	271.98	1602.92	

The cost of this system is on a cost per each window basis.

Bow/Bay Window System Components

Component Description	QUAN.	UNIT	LABOR-HOURS	COST EACH MAT.	COST EACH INST.	COST EACH TOTAL
Windows						
Windows, bow awning type, builder's quality, 8′ x 5′, insulating glass	1.000	Ea.	1.600	960.00	55.50	1015.50
Low E glass	1.000	Ea.	1.600	1225.00	55.50	1280.50
Plastic clad premium insulating glass, 6′ x 4′	1.000	Ea.	1.600	1050.00	55.50	1105.50
Metal clad deluxe insulating glass, 6′ x 4′	1.000	Ea.	1.600	865.00	55.50	920.50
Bow casement type, builder's quality, 8′ x 5′, single glass	1.000	Ea.	1.600	1475.00	55.50	1530.50
Insulating glass	1.000	Ea.	1.600	1775.00	55.50	1830.50
Plastic clad premium insulating glass, 8′ x 5′	1.000	Ea.	1.600	1250.00	55.50	1305.50
Metal clad deluxe insulating glass, 8′ x 5′	1.000	Ea.	1.600	1300.00	55.50	1355.50
Bow, double hung type, builder's quality, 8′ x 4′, single glass	1.000	Ea.	1.600	1025.00	55.50	1080.50
Insulating glass	1.000	Ea.	1.600	1100.00	55.50	1155.50

Fixed Window Systems

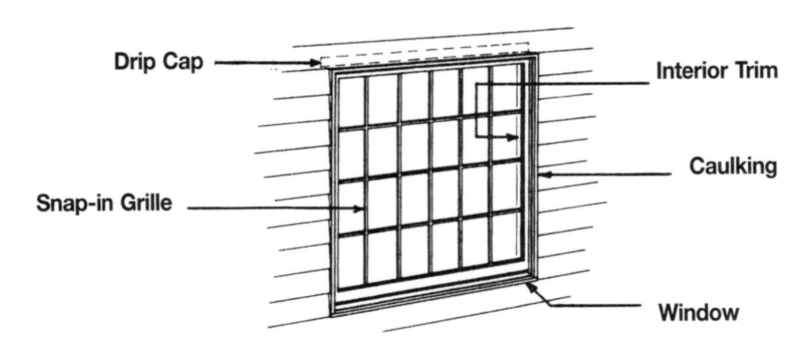

System Description	QUAN.	UNIT	LABOR-HOURS	COST EACH MAT.	COST EACH INST.	COST EACH TOTAL	____EA. EXTENSIONS
BUILDER'S QUALITY PICTURE WINDOW, 4' X 4'							
Window, primed, builder's quality, 4' x 4', insulating glass	1.000	Ea.	1.333	292.00	46.50	338.50	
Trim, interior casing	17.000	L.F.	.567	14.62	19.72	34.34	
Paint, interior & exterior, primer & 2 coats	2.000	Face	1.778	2.02	55.00	57.02	
Caulking	16.000	L.F.	.516	2.56	18.08	20.64	
Snap-in grille	1.000	Ea.	.267	132.00	9.30	141.30	
Drip cap, metal	4.000	L.F.	.080	.88	2.80	3.68	
TOTAL			4.541	444.08	151.40	595.48	
PLASTIC CLAD WOOD WINDOW, 4' X 4'							
Window, premium, 4' x 4', insulating glass	1.000	Ea.	1.333	475.00	46.50	521.50	
Trim, interior casing	17.000	L.F.	.567	14.62	19.72	34.34	
Paint, interior, primer & 2 coats	1.000	Face	.889	1.01	27.50	28.51	
Caulking	16.000	L.F.	.516	2.56	18.08	20.64	
Snap-in grille	1.000	Ea.	.267	132.00	9.30	141.30	
Drip cap, metal	4.000	L.F.	.080	.88	2.80	3.68	
TOTAL			3.652	626.07	123.90	749.97	
METAL CLAD WOOD WINDOW, 4' X 4'							
Window, deluxe, 4' x 4', insulating glass	1.000	Ea.	1.333	285.00	46.50	331.50	
Trim, interior casing	17.000	L.F.	.567	14.62	19.72	34.34	
Paint, interior, primer & 2 coats	1.000	Face	.889	1.01	27.50	28.51	
Caulking	16.000	L.F.	.516	2.56	18.08	20.64	
Snap-in grille	1.000	Ea.	.267	132.00	9.30	141.30	
Drip cap, metal	4.000	L.F.	.080	.88	2.80	3.68	
TOTAL			3.652	436.07	123.90	559.97	
BUILDER'S QUALITY PICTURE WINDOW, 6' X 4'-6''							
Window, builder's quality, 6' x 4'-6'', insulating glass	1.000	Ea.	1.600	505.00	55.50	560.50	
Trim, interior casing	23.000	L.F.	.767	19.78	26.68	46.46	
Paint, interior & exterior, primer & 2 coats	2.000	Face	1.778	2.02	55.00	57.02	
Caulking	22.000	L.F.	.710	3.52	24.86	28.38	
Snap-in grille	1.000	Ea.	.286	145.00	9.95	154.95	
Drip cap, metal	6.000	L.F.	.120	1.32	4.20	5.52	
TOTAL			5.261	676.64	176.19	852.83	

Fixed Window Systems

System Description	QUAN.	UNIT	LABOR-HOURS	COST EACH MAT.	COST EACH INST.	COST EACH TOTAL	____EA. EXTENSIONS
PLASTIC CLAD WOOD WINDOW, 4'-6" X 6'-6"							
Window, plastic clad, prem., 4'-6" x 6'-6", insul. glass	1.000	Ea.	1.455	650.00	50.50	700.50	
Trim, interior casing	23.000	L.F.	.767	19.78	26.68	46.46	
Paint, interior, primer & 2 coats	1.000	Face	.889	1.01	27.50	28.51	
Caulking	22.000	L.F.	.710	3.52	24.86	28.38	
Snap-in grille	1.000	Ea.	.267	132.00	9.30	141.30	
TOTAL			4.088	806.31	138.84	945.15	
METAL CLAD WOOD WINDOW, 4'-6" X 6'-6"							
Window, deluxe, 4'-6" x 6'-6", insulating glass	1.000	Ea.	1.455	420.00	50.50	470.50	
Trim, interior casing	23.000	L.F.	.767	19.78	26.68	46.46	
Paint, interior, primer & 2 coats	1.000	Face	.889	1.01	27.50	28.51	
Caulking	22.000	L.F.	.710	3.52	24.86	28.38	
Snap-in grille	1.000	Ea.	.267	132.00	9.30	141.30	
Drip cap, metal	4.500	L.F.	.090	.99	3.15	4.14	
TOTAL			4.178	577.30	141.99	719.29	
PLASTIC CLAD, WOOD, WINDOW, 6'-6" X 6'-6"							
Window, premium, 6'-6" x 6'-6", insulating glass	1.000	Ea.	1.600	865.00	55.50	920.50	
Trim, interior casing	27.000	L.F.	.900	23.22	31.32	54.54	
Paint, interior, primer & 2 coats	1.000	Face	1.600	5.40	49.50	54.90	
Caulking	26.000	L.F.	.839	4.16	29.38	33.54	
Snap-in grille	1.000	Ea.	.267	132.00	9.30	141.30	
Drip cap, metal	6.500	L.F.	.130	1.43	4.55	5.98	
TOTAL			5.336	1031.21	179.55	1210.76	
METAL CLAD WOOD WINDOW, 6'-6" X 6'-6"							
Window, metal clad, deluxe, 6'-6" x 6'-6", insulating glass	1.000	Ea.	1.600	530.00	55.50	585.50	
Trim interior casing	27.000	L.F.	.900	23.22	31.32	54.54	
Paint, interior, primer & 2 coats	1.000	Face	1.600	5.40	49.50	54.90	
Caulking	26.000	L.F.	.839	4.16	29.38	33.54	
Snap-in grille	1.000	Ea.	.267	132.00	9.30	141.30	
Drip cap, metal	6.500	L.F.	.130	1.43	4.55	5.98	
TOTAL			5.336	696.21	179.55	875.76	

The cost of this system is on a cost per each window basis.

Fixed Window System Components

Component Description	QUAN.	UNIT	LABOR-HOURS	COST EACH MAT.	COST EACH INST.	COST EACH TOTAL
Window, picture						
Window-picture, builder's quality, 4' x 4', single glass	1.000	Ea.	1.333	267.00	46.50	313.50
6' x 4'-6", single glass	1.000	Ea.	1.600	450.00	55.50	505.50
Plastic clad premium insulating glass, 4' x 4'	1.000	Ea.	1.333	475.00	46.50	521.50
5'-6" x 6'-6"	1.000	Ea.	1.600	845.00	55.50	900.50
Metal clad deluxe insulating glass, 4' x 4'	1.000	Ea.	1.333	285.00	46.50	331.50
5'-6" x 6'-6"	1.000	Ea.	1.600	465.00	55.50	520.50
Trim						
Trim, interior casing, window 4' x 4'	17.000	L.F.	.567	14.60	19.70	34.30
4'-6" x 4'-6"	19.000	L.F.	.633	16.35	22.00	38.35

Entrance Door Systems

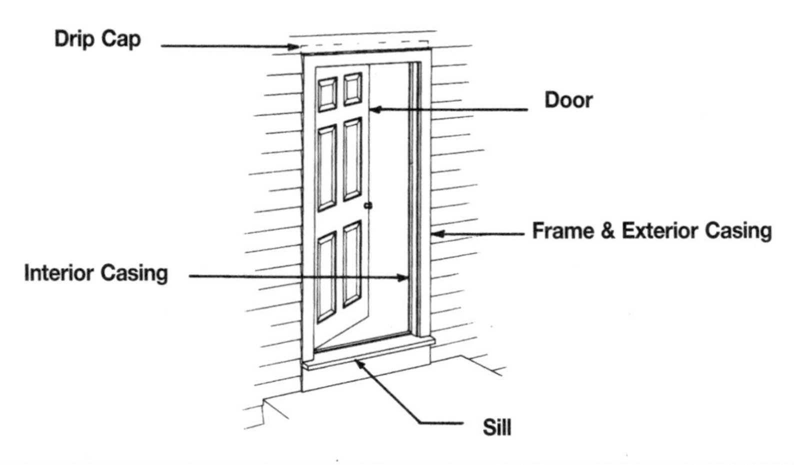

System Description	QUAN.	UNIT	LABOR-HOURS	COST EACH MAT.	COST EACH INST.	COST EACH TOTAL	____EA. EXTENSIONS
COLONIAL, 6 PANEL, 3' X 6'-8", WOOD							
Door, 3' x 6'-8" x 1-3/4" thick, pine, 6 panel colonial	1.000	Ea.	1.067	380.00	37.00	417.00	
Frame, 5-13/16" deep, incl. exterior casing & drip cap	17.000	L.F.	.725	124.95	25.16	150.11	
Interior casing, 2-1/2" wide	18.000	L.F.	.600	15.48	20.88	36.36	
Sill, 8/4 x 8" deep	3.000	L.F.	.480	43.95	16.65	60.60	
Butt hinges, brass, 4-1/2" x 4-1/2"	1.500	Pr.		13.43		13.43	
Lockset	1.000	Ea.	.571	31.00	19.90	50.90	
Weatherstripping, metal, spring type, bronze	1.000	Set	1.053	17.00	36.50	53.50	
Paint, interior & exterior, primer & 2 coats	2.000	Face	1.778	10.70	55.00	65.70	
TOTAL			6.274	636.51	211.09	847.60	
SOLID CORE BIRCH, FLUSH, 3' X 6'-8"							
Door, 3' x 6'-8", 1-3/4" thick, birch, flush solid core	1.000	Ea.	1.067	89.00	37.00	126.00	
Frame, 5-13/16" deep, incl. exterior casing & drip cap	17.000	L.F.	.725	124.95	25.16	150.11	
Interior casing, 2-1/2" wide	18.000	L.F.	.600	15.48	20.88	36.36	
Sill, 8/4 x 8" deep	3.000	L.F.	.480	43.95	16.65	60.60	
Butt hinges, brass, 4-1/2" x 4-1/2"	1.500	Pr.		13.43		13.43	
Lockset	1.000	Ea.	.571	31.00	19.90	50.90	
Weatherstripping, metal, spring type, bronze	1.000	Set	1.053	17.00	36.50	53.50	
Paint, interior & exterior, primer & 2 coats	2.000	Face	1.778	9.98	55.00	64.98	
TOTAL			6.274	344.79	211.09	555.88	
DUTCH, PINE, 3'-0" X 6'-8"							
Door, 3'-0" x 6'-8" x 1-3/4" thick, pine	1.000	Ea.	1.600	735.00	55.50	790.50	
Frame, 5-13/16" deep, exterior casing & drip cap	17.000	L.F.	.725	124.95	25.16	150.11	
Interior casing, 2-1/2" wide	19.000	L.F.	.600	15.48	20.88	36.36	
Sill, 8/4 x 8" deep	3.000	L.F.	.480	43.95	16.65	60.60	
Butt hinges, brass, 4-1/2" x 4-1/2"	1.500	Pr.		13.43		13.43	
Lockset	1.000	Ea.	.571	31.00	19.90	50.90	
Weatherstripping, metal, spring type, bronze	1.000	Set	1.053	17.00	36.50	53.50	
Paint, Interior & exterior, primer & 2 coats	2.000	Face	1.778	9.98	55.00	64.98	
TOTAL			6.807	990.79	229.59	1220.38	

Entrance Door Systems

System Description	QUAN.	UNIT	LABOR-HOURS	COST EACH MAT.	COST EACH INST.	COST EACH TOTAL	____EA. EXTENSIONS
COLONIAL, 8 PANEL, 3'-0" X 6'-8"							
Door, 3'-0" x 6'-8" x 1-3/4" thick, pine	1.000	Ea.	1.067	565.00	37.00	602.00	
Frame, 5-13/16" deep, exterior casing & drip cap	17.000	L.F.	.725	124.95	25.16	150.11	
Interior casing, 2-1/2" wide	19.000	L.F.	.600	15.48	20.88	36.36	
Sill, 8/4 x 8" deep	3.000	L.F.	.480	43.95	16.65	60.60	
Butt hinges, brass, 4-1/2" x 4-1/2"	1.500	Pr.		13.43		13.43	
Lockset	1.000	Ea.	.571	31.00	19.90	50.90	
Weatherstripping, metal, spring type, bronze	1.000	Set	1.053	17.00	36.50	53.50	
Paint, Interior & exterior, primer & 2 coats	2.000	Face	1.778	9.98	55.00	64.98	
TOTAL			6.274	820.79	211.09	1031.88	
METAL CLAD WOOD, RAISED PANEL, 3'-0" X 6'-8"							
Door, 3'-0" x 6'-8" x 1-3/8" thick, metal clad wood	1.000	Ea.	1.067	262.00	37.00	299.00	
Frame, 5-13/16" deep, exterior casing & drip cap	17.000	L.F.	.725	124.95	25.16	150.11	
Interior casing, 2-1/2" wide	19.000	L.F.	.600	15.48	20.88	36.36	
Sill, 8/4 x 8" deep	3.000	L.F.	.480	43.95	16.65	60.60	
Butt hinges, brass, 4-1/2" x 4-1/2"	1.500	Pr.		13.43		13.43	
Lockset	1.000	Ea.	.571	31.00	19.90	50.90	
Weatherstripping, metal, spring type, bronze	1.000	Set	1.053	17.00	36.50	53.50	
Paint, Interior & exterior, primer & 2 coats	2.000	Face	1.778	9.98	55.00	64.98	
TOTAL			6.274	517.79	211.09	728.88	
DELUXE METAL, RAISED PANEL, 3'-0" X 6'-8"							
Door, 3'-0" x 6'-8" x 1-3/8" thick, deluxe metal	1.000	Ea.	1.231	405.00	43.00	448.00	
Frame, 5-13/16" deep, exterior casing & drip cap	17.000	L.F.	.725	124.95	25.16	150.11	
Interior casing, 2-1/2" wide	19.000	L.F.	.600	15.48	20.88	36.36	
Sill, 8/4 x 8" deep	3.000	L.F.	.480	43.95	16.65	60.60	
Butt hinges, brass, 4-1/2" x 4-1/2"	1.500	Pr.		13.43		13.43	
Lockset	1.000	Ea.	.571	31.00	19.90	50.90	
Weatherstripping, metal, spring type, bronze	1.000	Set	1.053	17.00	36.50	53.50	
Paint, interior & exterior, primer & 2 coats	2.000	Face	1.778	9.98	55.00	64.98	
TOTAL			6.438	660.79	217.09	877.88	

These systems are on a cost per each door basis.

Entrance Door System Components

Component Description	QUAN.	UNIT	LABOR-HOURS	COST EACH MAT.	COST EACH INST.	COST EACH TOTAL
Doors, exterior						
Door exterior wood 1-3/4" thick, pine, dutch door, 2'-8" x 6'-8" minimum	1.000	Ea.	1.333	645.00	46.50	691.50
Maximum	1.000	Ea.	1.600	685.00	55.50	740.50
Hand carved mahogany, 2'-8" x 6'-8"	1.000	Ea.	1.067	470.00	37.00	507.00
3'-0" x 6'-8"	1.000	Ea.	1.067	505.00	37.00	542.00
Door, metal clad wood 1-3/8" thick raised panel, 2'-8" x 6'-8"	1.000	Ea.	1.067	264.00	37.00	301.00
Deluxe metal door, 2'-8" x 6'-8"	1.000	Ea.	1.231	410.00	43.00	453.00
Butt hinges						
Butt hinges, steel plated, 4-1/2" x 4-1/2", plain	1.500	Pr.		13.45		13.45
Ball bearing	1.500	Pr.		30.00		30.00
Lockset						
Lockset, minimum	1.000	Ea.	.571	31.00	19.90	50.90
Maximum	1.000	Ea.	1.000	130.00	35.00	165.00

Sliding Door Systems

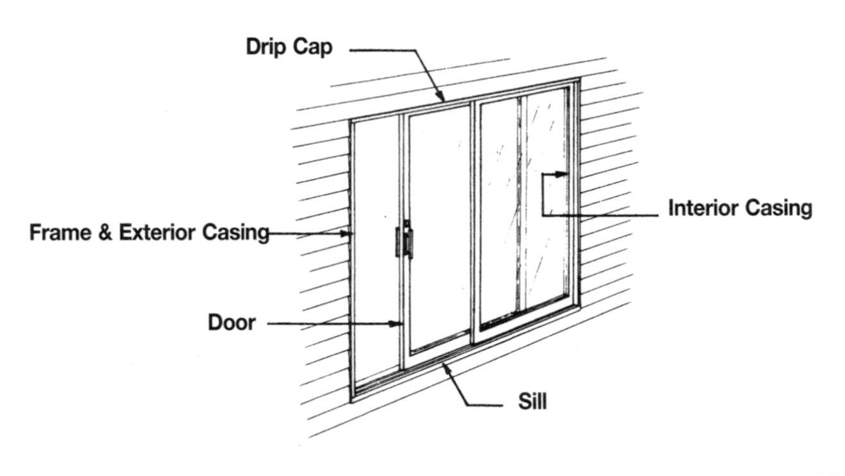

System Description	QUAN.	UNIT	LABOR-HOURS	COST EACH MAT.	COST EACH INST.	COST EACH TOTAL	___EA. EXTENSIONS
WOOD SLIDING DOOR, 6′ WIDE, PREMIUM							
Wood, 5/8″ thick tempered insul. glass, 6′ wide	1.000	Ea.	4.000	1025.00	139.00	1164.00	
Interior casing	20.000	L.F.	.667	17.20	23.20	40.40	
Exterior casing	20.000	L.F.	.667	17.20	23.20	40.40	
Sill, oak, 8/4 x 8″ deep	6.000	L.F.	.960	87.90	33.30	121.20	
Drip cap	6.000	L.F.	.120	1.32	4.20	5.52	
Paint, interior & exterior, primer & 2 coats	2.000	Face	2.560	13.60	79.20	92.80	
TOTAL			8.974	1162.22	302.10	1464.32	
WOOD SLIDING DOOR, 8′ WIDE, PREMIUM							
Wood, 5/8″ thick tempered insul. glass, 8′ wide, premium	1.000	Ea.	5.333	1200.00	186.00	1386.00	
Interior casing	22.000	L.F.	.733	18.92	25.52	44.44	
Exterior casing	22.000	L.F.	.733	18.92	25.52	44.44	
Sill, oak, 8/4 x 8″ deep	8.000	L.F.	1.280	117.20	44.40	161.60	
Drip cap	8.000	L.F.	.160	1.76	5.60	7.36	
Paint, interior & exterior, primer & 2 coats	2.000	Face	2.816	14.96	87.12	102.08	
TOTAL			11.055	1371.76	374.16	1745.92	
WOOD SLIDING DOOR, 12′ WIDE, PREMIUM							
Wood, 5/8″ thick tempered insul. glass, 12′ wide	1.000	Ea.	6.400	1925.00	223.00	2148.00	
Interior casing	26.000	L.F.	.867	22.36	30.16	52.52	
Exterior casing	26.000	L.F.	.867	22.36	30.16	52.52	
Sill, oak, 8/4 x 8″ deep	12.000	L.F.	1.920	175.80	66.60	242.40	
Drip cap	12.000	L.F.	.240	2.64	8.40	11.04	
Paint, interior & exterior, primer & 2 coats	2.000	Face	3.328	17.68	102.96	120.64	
TOTAL			13.622	2165.84	461.28	2627.12	
ALUMINUM SLIDING DOOR, 6′ WIDE, PREMIUM							
Aluminum, 5/8″ thick tempered insul. glass, 6′ wide	1.000	Ea.	4.000	1375.00	139.00	1514.00	
Interior casing	20.000	L.F.	.667	17.20	23.20	40.40	
Exterior casing	20.000	L.F.	.667	17.20	23.20	40.40	
Sill, oak, 8/4 x 8″ deep	6.000	L.F.	.960	87.90	33.30	121.20	
Drip cap	6.000	L.F.	.120	1.32	4.20	5.52	
Paint, interior & exterior, primer & 2 coats	2.000	Face	2.560	13.60	79.20	92.80	
TOTAL			8.974	1512.22	302.10	1814.32	

Sliding Door Systems

System Description	QUAN.	UNIT	LABOR-HOURS	COST EACH MAT.	COST EACH INST.	COST EACH TOTAL	EA. EXTENSIONS
ALUMINUM SLIDING DOOR, 8′ WIDE, PREMIUM							
Aluminum, 5/8″ tempered insul. glass, 8′ wide, premium	1.000	Ea.	5.333	1375.00	186.00	1561.00	
Interior casing	22.000	L.F.	.733	18.92	25.52	44.44	
Exterior casing	22.000	L.F.	.733	18.92	25.52	44.44	
Sill, oak, 8/4 x 8″ deep	8.000	L.F.	1.280	117.20	44.40	161.60	
Drip cap	8.000	L.F.	.160	1.76	5.60	7.36	
Paint, interior & exterior, primer & 2 coats	2.000	Face	2.816	14.96	87.12	102.08	
TOTAL			11.055	1546.76	374.16	1920.92	
ALUMINUM SLIDING DOOR, 12′ WIDE, PREMIUM							
Aluminum, 5/8″ thick tempered insul. glass, 12′ wide	1.000	Ea.	6.400	2275.00	223.00	2498.00	
Interior casing	26.000	L.F.	.867	22.36	30.16	52.52	
Exterior casing	26.000	L.F.	.867	22.36	30.16	52.52	
Sill, oak, 8/4 x 8″ deep	12.000	L.F.	1.920	175.80	66.60	242.40	
Drip cap	12.000	L.F.	.240	2.64	8.40	11.04	
Paint, interior & exterior, primer & 2 coats	2.000	Face	3.328	17.68	102.96	120.64	
TOTAL			13.622	2515.84	461.28	2977.12	

The cost of this system is on a cost per each door basis.

Sliding Door System Components

Component Description	QUAN.	UNIT	LABOR-HOURS	COST EACH MAT.	COST EACH INST.	COST EACH TOTAL
Sliding doors						
Sliding door, wood, 5/8″ thick, tempered insul. glass, 6′ wide, premium	1.000	Ea.	4.000	1025.00	139.00	1164.00
Economy	1.000	Ea.	4.000	745.00	139.00	884.00
8′wide, wood premium	1.000	Ea.	5.333	1200.00	186.00	1386.00
Economy	1.000	Ea.	5.333	875.00	186.00	1061.00
12′ wide, wood premium	1.000	Ea.	6.400	1925.00	223.00	2148.00
Economy	1.000	Ea.	6.400	1250.00	223.00	1473.00
Aluminum, 5/8″ thick, tempered insul. glass, 6′wide, premium	1.000	Ea.	4.000	1375.00	139.00	1514.00
Economy	1.000	Ea.	4.000	565.00	139.00	704.00
8′wide, premium	1.000	Ea.	5.333	1375.00	186.00	1561.00
Economy	1.000	Ea.	5.333	1150.00	186.00	1336.00
12′ wide, premium	1.000	Ea.	6.400	2275.00	223.00	2498.00
Economy	1.000	Ea.	6.400	1425.00	223.00	1648.00
Sills						
Sill, oak, 8/4 x 8″ deep, 6′ wide door	6.000	L.F.	.960	88.00	33.50	121.50
8′ wide door	8.000	L.F.	1.280	117.00	44.50	161.50
12′ wide door	12.000	L.F.	1.920	176.00	66.50	242.50
8/4 x 10″ deep, 6′ wide door	6.000	L.F.	1.067	119.00	37.00	156.00
8′ wide door	8.000	L.F.	1.422	158.00	49.50	207.50
12′ wide door	12.000	L.F.	2.133	237.00	74.50	311.50
Paint or Stain						
Paint or stain, interior & exterior, 6′ wide door, 1 coat	2.000	Face	1.600	4.80	49.50	54.30
8′ wide door, 1 coat	2.000	Face	1.760	5.30	54.50	59.80
12′ wide door, 1 coat	2.000	Face	2.080	6.25	64.50	70.75
Aluminum door, trim only, interior & exterior, 6′ door, 1 coat	2.000	Face	.800	2.40	25.00	27.40
8′ wide door, 1 coat	2.000	Face	.880	2.64	27.50	30.14
12′ wide door, 1 coat	2.000	Face	1.040	3.12	32.00	35.12

Residential Garage Door Systems

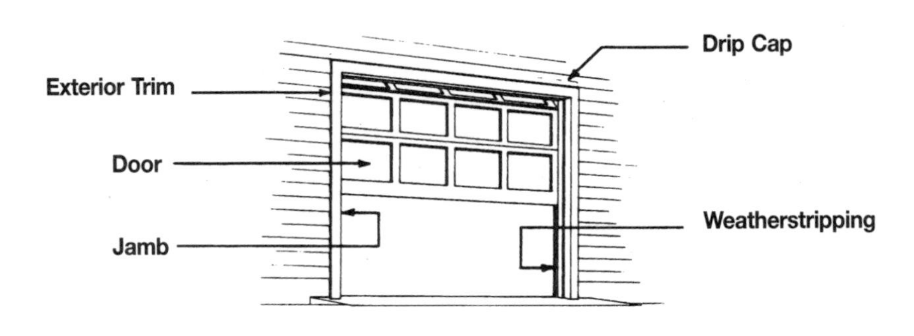

System Description	QUAN.	UNIT	LABOR-HOURS	COST EACH MAT.	COST EACH INST.	COST EACH TOTAL	____EA. EXTENSIONS
OVERHEAD, SECTIONAL GARAGE DOOR, 9′ X 7′							
Wood, overhead sectional door, std., incl. hardware, 9′ x 7′	1.000	Ea.	2.000	400.00	69.50	469.50	
Jamb & header blocking, 2″ x 6″	25.000	L.F.	.901	16.25	31.25	47.50	
Exterior trim	25.000	L.F.	.833	21.50	29.00	50.50	
Paint, interior & exterior, primer & 2 coats	2.000	Face	3.556	21.40	110.00	131.40	
Weatherstripping, molding type	1.000	Set	.767	19.78	26.68	46.46	
Drip cap	9.000	L.F.	.180	1.98	6.30	8.28	
TOTAL			8.237	480.91	272.73	753.64	
OVERHEAD, METAL, SECTIONAL GARAGE DOOR, 9′ X 7′							
Metal, overhead door, standard, incl. hardware, 9′ x 7′	1.000	Ea.	3.030	285.00	105.00	390.00	
Jamb & header blocking, 2″ x 6″	25.000	L.F.	.901	16.25	31.25	47.50	
Exterior trim	25.000	L.F.	.833	21.50	29.00	50.50	
Paint, interior & exterior, primer & 2 coats	2.000	Face	3.556	21.40	110.00	131.40	
Weatherstripping, molding type	1.000	Set	.767	19.78	26.68	46.46	
Drip cap	9.000	L.F.	.180	1.98	6.30	8.28	
TOTAL			9.267	365.91	308.23	674.14	
OVERHEAD, SWING-UP TYPE, GARAGE DOOR, 9′ X 7′							
Wood, overhead door, standard, incl. hardware, 9′ x 7′	1.000	Ea.	2.000	345.00	69.50	414.50	
Jamb & header blocking, 2″ x 6″	25.000	L.F.	.901	16.25	31.25	47.50	
Exterior trim	25.000	L.F.	.833	21.50	29.00	50.50	
Paint, interior & exterior, primer & 2 coats	2.000	Face	3.556	21.40	110.00	131.40	
Weatherstripping, molding type	1.000	Set	.767	19.78	26.68	46.46	
Drip cap	9.000	L.F.	.180	1.98	6.30	8.28	
TOTAL			8.237	425.91	272.73	698.64	
OVERHEAD, SECTIONAL GARAGE DOOR, 16′ X 7′							
Wood, overhead sectional, std., incl. hardware, 16′ x 7′	1.000	Ea.	2.667	805.00	93.00	898.00	
Jamb & header blocking, 2″ x 6″	30.000	L.F.	1.081	19.50	37.50	57.00	
Exterior trim	30.000	L.F.	1.000	25.80	34.80	60.60	
Paint, interior & exterior, primer & 2 coats	2.000	Face	5.333	32.10	165.00	197.10	
Weatherstripping, molding type	1.000	Set	1.000	25.80	34.80	60.60	
Drip cap	16.000	L.F.	.320	3.52	11.20	14.72	
TOTAL			11.401	911.72	376.30	1288.02	

Residential Garage Door Systems

System Description	QUAN.	UNIT	LABOR-HOURS	COST EACH MAT.	COST EACH INST.	COST EACH TOTAL	____EA. EXTENSIONS
OVERHEAD, METAL, SECTIONAL GARAGE DOOR, 16′ X 7′							
Metal, overhead door, standard, incl. hardware 16′ x 7′	1.000	Ea.	5.333	560.00	186.00	746.00	
Jamb & header blocking, 2″ x 6″	30.000	L.F.	1.081	19.50	37.50	57.00	
Exterior trim	30.000	L.F.	1.000	25.80	34.80	60.60	
Paint, interior & exterior, primer & 2 coats	2.000	Face	5.333	32.10	165.00	197.10	
Weatherstripping, molding type	1.000	Set	1.000	25.80	34.80	60.60	
Drip cap	16.000	L.F.	.320	3.52	11.20	14.72	
TOTAL			14.067	666.72	469.30	1136.02	
OVERHEAD, SWING-UP TYPE, GARAGE DOOR, 16′ X 7′							
Wood, overhead, swing-up, std., incl. hardware, 16′ x 7′	1.000	Ea.	2.667	600.00	93.00	693.00	
Jamb & header blocking, 2″ x 6″	30.000	L.F.	1.081	19.50	37.50	57.00	
Exterior trim	30.000	L.F.	1.000	25.80	34.80	60.60	
Paint, interior & exterior, primer & 2 coats	2.000	Face	5.333	32.10	165.00	197.10	
Weatherstripping, molding type	1.000	Set	1.000	25.80	34.80	60.60	
Drip cap	16.000	L.F.	.320	3.52	11.20	14.72	
TOTAL			11.401	706.72	376.30	1083.02	

This system is on a cost per each door basis.

Residential Garage Door System Components

Component Description	QUAN.	UNIT	LABOR-HOURS	COST EACH MAT.	COST EACH INST.	COST EACH TOTAL
Garage doors						
Overhead, sectional, including hardware, fiberglass, 9′ x 7′, standard	1.000	Ea.	3.030	565.00	105.00	670.00
Deluxe	1.000	Ea.	3.030	660.00	105.00	765.00
16′ x 7′, standard	1.000	Ea.	2.667	875.00	93.00	968.00
Deluxe	1.000	Ea.	2.667	1100.00	93.00	1193.00
Hardboard, 9′ x 7′, standard	1.000	Ea.	2.000	360.00	69.50	429.50
Deluxe	1.000	Ea.	2.000	435.00	69.50	504.50
16′ x 7′, standard	1.000	Ea.	2.667	635.00	93.00	728.00
Deluxe	1.000	Ea.	2.667	740.00	93.00	833.00
Overhead swing-up type including hardware, fiberglass, 9′ x 7′, standard	1.000	Ea.	2.000	555.00	69.50	624.50
Deluxe	1.000	Ea.	2.000	660.00	69.50	729.50
16′ x 7′, standard	1.000	Ea.	2.667	700.00	93.00	793.00
Deluxe	1.000	Ea.	2.667	815.00	93.00	908.00
Hardboard, 9′ x 7′, standard	1.000	Ea.	2.000	287.00	69.50	356.50
Deluxe	1.000	Ea.	2.000	380.00	69.50	449.50
16′ x 7′, standard	1.000	Ea.	2.667	400.00	93.00	493.00
Deluxe	1.000	Ea.	2.667	595.00	93.00	688.00
Metal, 9′ x 7′, standard	1.000	Ea.	2.000	315.00	69.50	384.50
Deluxe	1.000	Ea.	2.000	490.00	69.50	559.50
16′ x 7′, standard	1.000	Ea.	2.667	490.00	93.00	583.00
Deluxe	1.000	Ea.	2.667	795.00	93.00	888.00
Jamb and header blocking						
Jamb & header blocking, 2″ x 6″, 9′ x 7′ door	25.000	L.F.	.901	16.25	31.50	47.75
2″ x 8″, 9′ x 7′ door	25.000	L.F.	1.000	23.00	35.00	58.00
16′ x 7′ door	30.000	L.F.	1.200	27.50	41.50	69.00
Garage door openers						
Garage door opener, economy	1.000	Ea.	1.000	231.00	35.00	266.00
Deluxe, including remote control	1.000	Ea.	1.000	330.00	35.00	365.00

Aluminum Window Systems

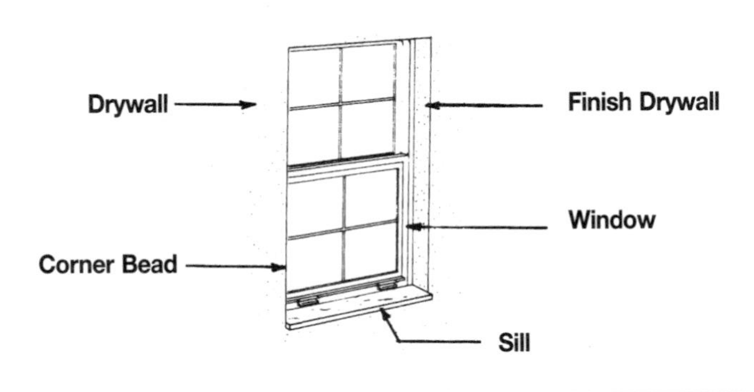

System Description	QUAN.	UNIT	LABOR-HOURS	COST EACH MAT.	COST EACH INST.	COST EACH TOTAL	EA. EXTENSIONS
SINGLE HUNG, 2′ X 3′ OPENING							
Window, 2′ x 3′ opening, enameled, insulating glass	1.000	Ea.	1.600	168.00	69.00	237.00	
Blocking, 1″ x 3″ furring strip nailers	10.000	L.F.	.146	2.10	5.10	7.20	
Drywall, 1/2″ thick, standard	5.000	S.F.	.040	1.45	1.40	2.85	
Corner bead, 1″ x 1″, galvanized steel	8.000	L.F.	.160	.88	5.60	6.48	
Finish drywall, tape and finish corners inside and outside	16.000	L.F.	.233	1.12	8.16	9.28	
Sill, slate	2.000	L.F.	.400	16.00	12.20	28.20	
TOTAL			2.579	189.55	101.46	291.01	
SLIDING, 3′ X 2′ OPENING							
Window, 3′ x 2′ opening, enameled, insulating glass	1.000	Ea.	1.600	173.00	69.00	242.00	
Blocking, 1″ x 3″ furring strip nailers	10.000	L.F.	.146	2.10	5.10	7.20	
Drywall, 1/2″ thick, standard	5.000	S.F.	.040	1.45	1.40	2.85	
Corner bead, 1″ x 1″, galvanized steel	7.000	L.F.	.140	.77	4.90	5.67	
Finish drywall, tape and finish corners inside and outside	14.000	L.F.	.204	.98	7.14	8.12	
Sill, slate	3.000	L.F.	.600	24.00	18.30	42.30	
TOTAL			2.730	202.30	105.84	308.14	
AWNING, 3′-1″ X 3′-2″							
Window, 3′-1″ x 3′-2″ opening, enameled, insul. glass	1.000	Ea.	1.600	193.00	69.00	262.00	
Blocking, 1″ x 3″ furring strip, nailers	12.500	L.F.	.182	2.63	6.38	9.01	
Drywall, 1/2″ thick, standard	4.500	S.F.	.036	1.31	1.26	2.57	
Corner bead, 1″ x 1″, galvanized steel	9.250	L.F.	.185	1.02	6.48	7.50	
Finish drywall, tape and finish corners, inside and outside	18.500	L.F.	.269	1.30	9.44	10.74	
Sill, slate	3.250	L.F.	.650	26.00	19.83	45.83	
TOTAL			2.922	225.26	112.39	337.65	
CASEMENT, 3′-1″ X 3′-2″							
Window, 3′-1″ x 3′-2″ opening, enameled, insul. glass	1.000	Ea.	1.600	289.00	69.00	358.00	
Blocking, 1″ x 3″ furring strip nailers	12.500	L.F.	.182	2.63	6.38	9.01	
Drywall, 1/2″ thick standard	6.000	S.F.	.048	1.74	1.68	3.42	
Corner bead, 1″ x 1″ galvanized steel	9.000	L.F.	.180	.99	6.30	7.29	
Finish drywall, tape & finish corners, inside & outside	18.000	L.F.	.262	1.26	9.18	10.44	
Sill, slate	3.000	L.F.	.600	24.00	18.30	42.30	
TOTAL			2.872	319.62	110.84	430.46	

Aluminum Window Systems

System Description	QUAN.	UNIT	LABOR-HOURS	COST EACH MAT.	COST EACH INST.	COST EACH TOTAL	___EA. EXTENSIONS
SINGLE HUNG, 2'-8" X 6'-8"							
Window, 2'-8" x 6'-8" opening, enameled, insul. glass	1.000	Ea.	2.000	380.00	86.00	466.00	
Blocking, 1" x 3" furring strip, nailers	19.000	L.F.	.276	3.99	9.69	13.68	
Drywall, 1/2" thick standard	9.500	S.F.	.076	2.76	2.66	5.42	
Corner bead, 1" x 1" galvanized steel	15.000	L.F.	.300	1.65	10.50	12.15	
Finish drywall, tape & finish corners, inside & outside	30.000	L.F.	.437	2.10	15.30	17.40	
Sill, slate	3.000	L.F.	.600	24.00	18.30	42.30	
TOTAL			3.689	414.50	142.45	556.95	
SLIDING, 5' X 3'							
Window, 5' x 3' opening, enameled, insulating glass	1.000	Ea.	1.778	278.00	76.50	354.50	
Blocking, 1" x 3" furring strip, nailers	16.000	L.F.	.233	3.36	8.16	11.52	
Drywall, 1/2" thick, standard	8.000	L.F.	.064	2.32	2.24	4.56	
Corner bead, 1" x 1" galvanized steel	11.000	L.F.	.220	1.21	7.70	8.91	
Finish drywall, tape & finish corners, inside & outside	22.000	L.F.	.320	1.54	11.22	12.76	
Sill, slate	3.000	L.F.	.600	24.00	18.30	42.30	
TOTAL			3.215	310.43	124.12	434.55	
AWNING, 4'-5" X 5'-3"							
Window, 4'-5" x 5'-3" opening, enameled, insul. glass	1.000	Ea.	2.000	335.00	86.00	421.00	
Blocking, 1" x 3" furring strip, nailers	20.000	L.F.	.262	3.78	9.18	12.96	
Drywall, 1/2" thick, standard	10.000	S.F.	.072	2.61	2.52	5.13	
Corner bead, 1" x 1" galvanized steel	14.000	L.F.	.260	1.43	9.10	10.53	
Finish drywall, tape & finish corners, inside & outside	28.000	L.F.	.378	1.82	13.26	15.08	
Sill, slate	4.500	L.F.	.800	32.00	24.40	56.40	
TOTAL			3.772	376.64	144.46	521.10	
SINGLE HUNG, 3'-4" X 5'-0"							
Window, 3'-4" x 5'-0" opening, enameled, insul. glass	1.000	Ea.	1.778	190.00	76.50	266.50	
Blocking, 1" x 3" furring strip, nailers	16.000	L.F.	.233	3.36	8.16	11.52	
Drywall, 1/2" thick standard	8.000	S.F.	.064	2.32	2.24	4.56	
Corner bead, 1" x 1" galvanized steel	11.000	L.F.	.220	1.21	7.70	8.91	
Finish drywall, tape & finish corners, inside & outside	22.000	L.F.	.320	1.54	11.22	12.76	
Sill, slate	3.500	L.F.	.600	24.00	18.30	42.30	
TOTAL			3.215	222.43	124.12	346.55	

Aluminum Window System Components

Component Description	QUAN.	UNIT	LABOR-HOURS	COST EACH MAT.	COST EACH INST.	COST EACH TOTAL
Windows, aluminum						
Window, aluminum, awning, 3'-1" x 3'-2", standard glass	1.000	Ea.	1.600	207.00	69.00	276.00
4'-5" x 5'-3", standard glass	1.000	Ea.	2.000	291.00	86.00	377.00
Casement, 3'-1" x 3'-2", standard glass	1.000	Ea.	1.600	289.00	69.00	358.00
Single hung, 2' x 3', standard glass	1.000	Ea.	1.600	139.00	69.00	208.00
3'-4" x 5'-0", standard glass	1.000	Ea.	1.778	190.00	76.50	266.50
Sliding, 3' x 2', standard glass	1.000	Ea.	1.600	156.00	69.00	225.00
8' x 4', standard glass	1.000	Ea.	2.667	286.00	115.00	401.00
Insulating glass	1.000	Ea.	2.667	460.00	115.00	575.00
Sills						
Wood, 1-5/8" x 5-1/8", 2' long	2.000	L.F.	.128	5.20	4.46	9.66
3' long	3.000	L.F.	.192	7.75	6.70	14.45
4' long	4.000	L.F.	.256	10.35	8.90	19.25

Storm Door & Window Systems

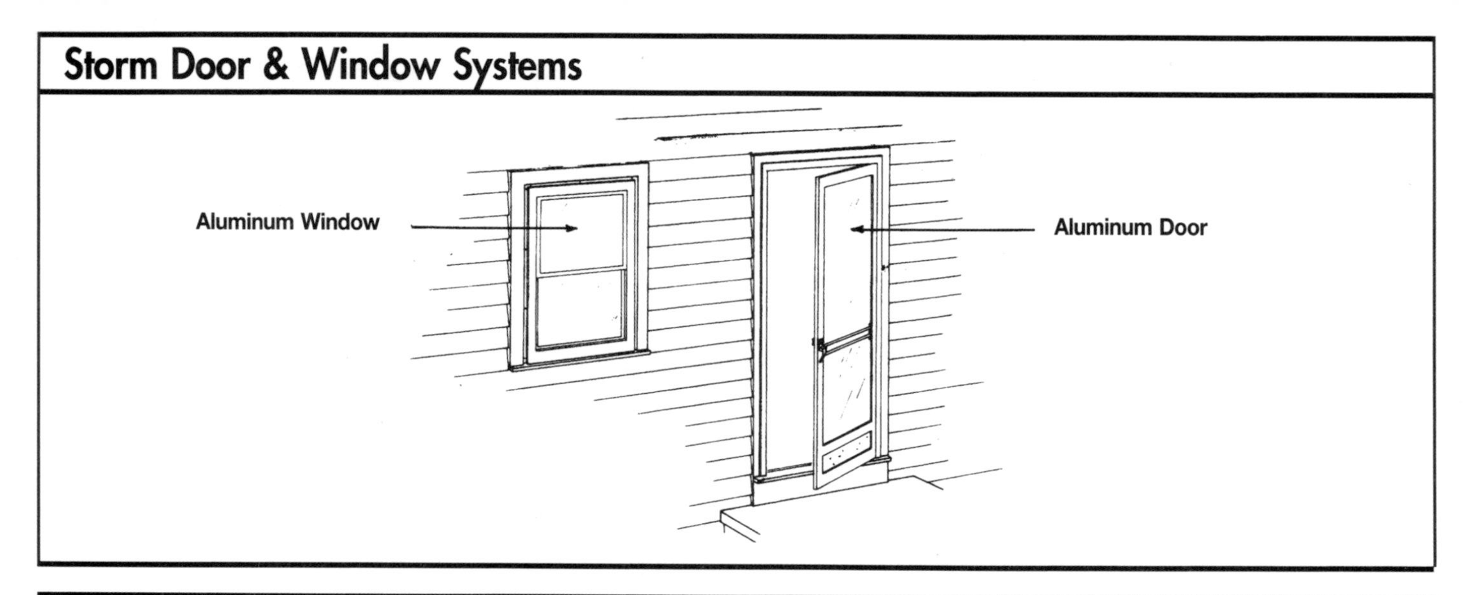

Component Description	QUAN.	UNIT	LABOR-HOURS	COST EACH		
				MAT.	INST.	TOTAL
Storm doors						
Storm door, aluminum, combination, storm & screen, anodized, 2'-6" x 6'-8"	1.000	Ea.	1.067	153.00	37.00	190.00
2'-8" x 6'-8"	1.000	Ea.	1.143	183.00	40.00	223.00
3'-0" x 6'-8"	1.000	Ea.	1.143	183.00	40.00	223.00
Mill finish, 2'-6" x 6'-8"	1.000	Ea.	1.067	215.00	37.00	252.00
2'-8" x 6'-8"	1.000	Ea.	1.143	215.00	40.00	255.00
3'-0" x 6'-8"	1.000	Ea.	1.143	233.00	40.00	273.00
Painted, 2'-6" x 6'-8"	1.000	Ea.	1.067	210.00	37.00	247.00
2'-8" x 6'-8"	1.000	Ea.	1.143	218.00	40.00	258.00
3'-0" x 6'-8"	1.000	Ea.	1.143	225.00	40.00	265.00
Wood, combination, storm & screen, crossbuck, 2'-6" x 6'-9"	1.000	Ea.	1.455	239.00	50.50	289.50
2'-8" x 6'-9"	1.000	Ea.	1.600	242.00	55.50	297.50
3'-0" x 6'-9"	1.000	Ea.	1.778	251.00	62.00	313.00
Full lite, 2'-6" x 6'-9"	1.000	Ea.	1.455	242.00	50.50	292.50
2'-8" x 6'-9"	1.000	Ea.	1.600	242.00	55.50	297.50
3'-0" x 6'-9"	1.000	Ea.	1.778	251.00	62.00	313.00
Windows						
Windows, aluminum, combination storm & screen, basement, 1'-10" x 1'-0"	1.000	Ea.	.533	28.00	18.55	46.55
2'-9" x 1'-6"	1.000	Ea.	.533	30.50	18.55	49.05
3'-4" x 2'-0"	1.000	Ea.	.533	36.50	18.55	55.05
Double hung, anodized, 2'-0" x 3'-5"	1.000	Ea.	.533	72.00	18.55	90.55
2'-6" x 5'-0"	1.000	Ea.	.571	96.50	19.90	116.40
4'-0" x 6'-0"	1.000	Ea.	.640	204.00	22.50	226.50
Painted, 2'-0" x 3'-5"	1.000	Ea.	.533	86.00	18.55	104.55
2'-6" x 5'-0"	1.000	Ea.	.571	138.00	19.90	157.90
4'-0" x 6'-0"	1.000	Ea.	.640	247.00	22.50	269.50
Fixed window, anodized, 4'-6" x 4'-6"	1.000	Ea.	.640	110.00	22.50	132.50
5'-8" x 4'-6"	1.000	Ea.	.800	125.00	28.00	153.00
Painted, 4'-6" x 4'-6"	1.000	Ea.	.640	110.00	22.50	132.50
5'-8" x 4'-6"	1.000	Ea.	.800	125.00	28.00	153.00

Shutters/Blinds Systems

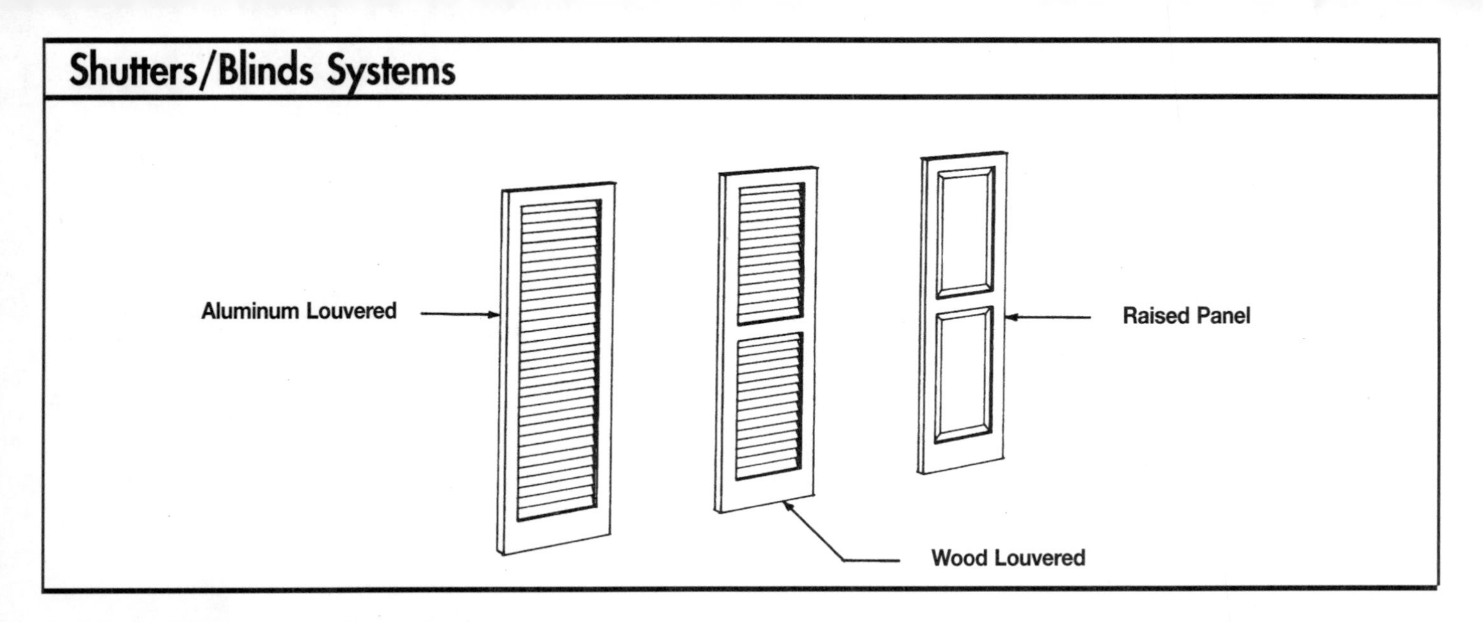

Component Description	QUAN.	UNIT	LABOR-HOURS	COST EACH		
				MAT.	INST.	TOTAL
Shutters, exterior blinds						
Shutters, exterior blinds, aluminum, louvered, 1'-4" wide, 3"-0" long	1.000	Set	.800	43.50	28.00	71.50
4'-0" long	1.000	Set	.800	52.00	28.00	80.00
5'-4" long	1.000	Set	.800	68.50	28.00	96.50
6'-8" long	1.000	Set	.889	87.00	31.00	118.00
Wood, louvered, 1'-2" wide, 3'-3" long	1.000	Set	.800	82.50	28.00	110.50
4'-7" long	1.000	Set	.800	111.00	28.00	139.00
5'-3" long	1.000	Set	.800	126.00	28.00	154.00
1'-6" wide, 3'-3" long	1.000	Set	.800	87.50	28.00	115.50
4'-7" long	1.000	Set	.800	123.00	28.00	151.00
Polystyrene, solid raised panel, 3'-3" wide, 3'-0" long	1.000	Set	.800	161.00	28.00	189.00
3'-11" long	1.000	Set	.800	176.00	28.00	204.00
5'-3" long	1.000	Set	.800	231.00	28.00	259.00
6'-8" long	1.000	Set	.889	260.00	31.00	291.00
Polystyrene, louvered, 1'-2" wide, 3'-3" long	1.000	Set	.800	44.00	28.00	72.00
4'-7" long	1.000	Set	.800	54.50	28.00	82.50
5'-3" long	1.000	Set	.800	58.50	28.00	86.50
6'-8" long	1.000	Set	.889	95.50	31.00	126.50
Vinyl, louvered, 1'-2" wide, 4'-7" long	1.000	Set	.720	51.50	25.00	76.50
1'-4" x 6'-8" long	1.000	Set	.889	87.00	31.00	118.00

Roofing Systems

Gable End Roofing Systems

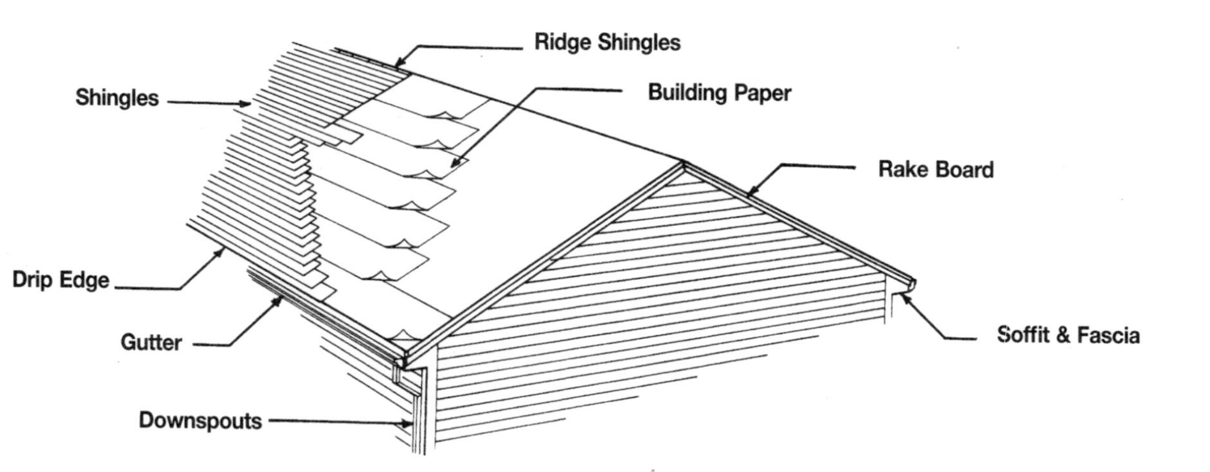

System Description	QUAN.	UNIT	LABOR-HOURS	COST PER S.F. MAT.	COST PER S.F. INST.	COST PER S.F. TOTAL	___S.F. EXTENSIONS
ASPHALT, ROOF SHINGLES, CLASS A							
Shingles, inorganic class A, 210-235 lb./sq., 4/12 pitch	1.160	S.F.	.017	.36	.58	.94	
Drip edge, metal, 5″ wide	.150	L.F.	.003	.04	.11	.15	
Building paper, #15 felt	1.300	S.F.	.002	.03	.06	.09	
Ridge shingles, asphalt	.042	L.F.	.001	.03	.03	.06	
Soffit & fascia, white painted aluminum, 1′ overhang	.083	L.F.	.012	.20	.42	.62	
Rake trim, painted, 1″ x 6″	.040	L.F.	.004	.04	.12	.16	
Gutter, seamless, aluminum painted	.083	L.F.	.006	.10	.21	.31	
Downspouts, aluminum painted	.035	L.F.	.002	.04	.06	.10	
TOTAL			.047	.84	1.59	2.43	
ASPHALT, ROOF SHINGLES, PREMIUM LAMINATED							
Shingles, multi-layered, 260-300 lb./sq., 4/12 pitch	1.160	S.F.	.027	.64	.91	1.55	
Drip edge, metal, 5″ wide	.150	L.F.	.003	.04	.11	.15	
Building paper, #15 felt	1.300	S.F.	.002	.03	.06	.09	
Ridge shingles, asphalt	.042	L.F.	.001	.03	.03	.06	
Rake trim, painted, 1″ x 6″	.040	L.F.	.004	.04	.12	.16	
Soffit & fascia, white painted aluminum, 1′ overhang	.083	L.F.	.012	.20	.42	.62	
Gutter, seamless, aluminum painted	.083	L.F.	.006	.10	.21	.31	
Downspouts, aluminum painted	.035	L.F.	.002	.04	.06	.10	
TOTAL			.057	1.12	1.92	3.04	
CLAY TILE, SPANISH TILE, RED							
Clay tile, Spanish tile, red, 4/12 pitch	1.160	S.F.	.053	3.96	1.78	5.74	
Drip edge, metal, 5″ wide	.150	L.F.	.003	.04	.11	.15	
Building paper, #15 felt	1.300	S.F.	.002	.03	.06	.09	
Ridge tiles, clay	.042	L.F.	.002	.40	.06	.46	
Soffit & fascia, white painted aluminum, 1′ overhang	.083	L.F.	.012	.20	.42	.62	
Rake trim, painted, 1″ x 6″	.040	L.F.	.004	.04	.12	.16	
Gutter, seamless, aluminum painted	.083	L.F.	.006	.10	.21	.31	
Downspouts, aluminum painted	.035	L.F.	.002	.04	.06	.10	
TOTAL			.084	4.81	2.82	7.63	

Gable End Roofing Systems

System Description	QUAN.	UNIT	LABOR-HOURS	COST PER S.F.			S.F. EXTENSIONS
				MAT.	INST.	TOTAL	
WOOD, CEDAR SHINGLES NO. 1 PERFECTIONS, 18″ LONG							
Shingles, wood, cedar, No. 1 perfections, 4/12 pitch	1.160	S.F.	.035	2.11	1.21	3.32	
Drip edge, metal, 5″ wide	.150	L.F.	.003	.04	.11	.15	
Building paper, #15 felt	1.300	S.F.	.002	.03	.06	.09	
Ridge shingles, cedar	.042	L.F.	.001	.11	.04	.15	
Soffit & fascia, white painted aluminum, 1′ overhang	.083	L.F.	.012	.20	.42	.62	
Rake trim, painted, 1″ x 6″	.040	L.F.	.004	.04	.12	.16	
Gutter, seamless, aluminum, painted	.083	L.F.	.006	.10	.21	.31	
Downspouts, aluminum, painted	.035	L.F.	.002	.04	.06	.10	
TOTAL			.065	2.67	2.23	4.90	
WOOD, CEDAR SHINGLES, RESQUARED AND REBUTTED, 18″ LONG							
Shingles, Wood, cedar, resquared & rebutted, 4/12 pitch	1.160	S.F.	.032	2.63	1.12	3.75	
Drip edge, metal, 5″ wide	.150	L.F.	.003	.04	.11	.15	
Building paper, #15 felt	1.300	S.F.	.002	.03	.06	.09	
Ridge shingles, cedar	.042	L.F.	.001	.11	.04	.15	
Soffit & fascia, white painted aluminum, 1′ overhang	.083	L.F.	.012	.20	.42	.62	
Rake trim, painted, 1″ x 6″	.040	L.F.	.004	.04	.12	.16	
Gutter, seamless, aluminum, painted	.083	L.F.	.006	.10	.21	.31	
Downspouts, aluminum, painted	.035	L.F.	.002	.04	.06	.10	
TOTAL			.062	3.19	2.14	5.33	
WOOD SHAKES							
Shingles, hand split, 24″ long, 10″ exposure 4/12 pitch	1.160	S.F.	.038	1.82	1.33	3.15	
Drip edge, metal, 5″ wide	.150	L.F.	.003	.04	.11	.15	
Building paper, #15 felt	1.300	S.F.	.002	.03	.06	.09	
Ridge shingles, shakes	.042	L.F.	.001	.11	.04	.15	
Soffit & fascia, white painted aluminum, 1′ overhang	.083	L.F.	.012	.20	.42	.62	
Rake trim, painted, 1″ x 6″	.040	L.F.	.004	.04	.12	.16	
Gutter, seamless, aluminum, painted	.083	L.F.	.006	.10	.21	.31	
Downspouts, aluminum, painted	.035	L.F.	.002	.04	.06	.10	
TOTAL			.068	2.38	2.35	4.73	

The prices in these systems are based on a square foot of plan area.
All quantities have been adjusted accordingly.

Gable End Roofing System Components

Component Description	QUAN.	UNIT	LABOR-HOURS	COST PER S.F.		
				MAT.	INST.	TOTAL
Shingles						
Shingles, asphalt, inorganic, class A, 210-235 lb./sq., 4/12 pitch	1.160	S.F.	.017	.36	.58	.94
8/12 pitch	1.330	S.F.	.019	.39	.62	1.01
Mission tile, red, 4/12 pitch	1.160	S.F.	.083	8.35	2.78	11.13
8/12 pitch	1.330	S.F.	.090	9.05	3.02	12.07
Slate, Buckingham, Virginia, black, 4/12 pitch	1.160	S.F.	.055	7.15	1.82	8.97
8/12 pitch	1.330	S.F.	.059	7.75	1.98	9.73
Wood, No. 1 red cedar, 5X, 16″ long, 5″ exposure, 4/12 pitch	1.160	S.F.	.038	1.99	1.33	3.32
8/12 pitch	1.330	S.F.	.042	2.16	1.44	3.60
Fire retardant, 4/12 pitch	1.160	S.F.	.038	2.39	1.33	3.72
8/12 pitch	1.330	S.F.	.042	2.59	1.44	4.03
Wood shakes hand split, 24″ long, 10″ exposure, 4/12 pitch	1.160	S.F.	.038	1.82	1.33	3.15
18″ long, 8″ exposure, 4/12 pitch	1.160	S.F.	.048	1.28	1.67	2.95

Hip Roof Roofing Systems

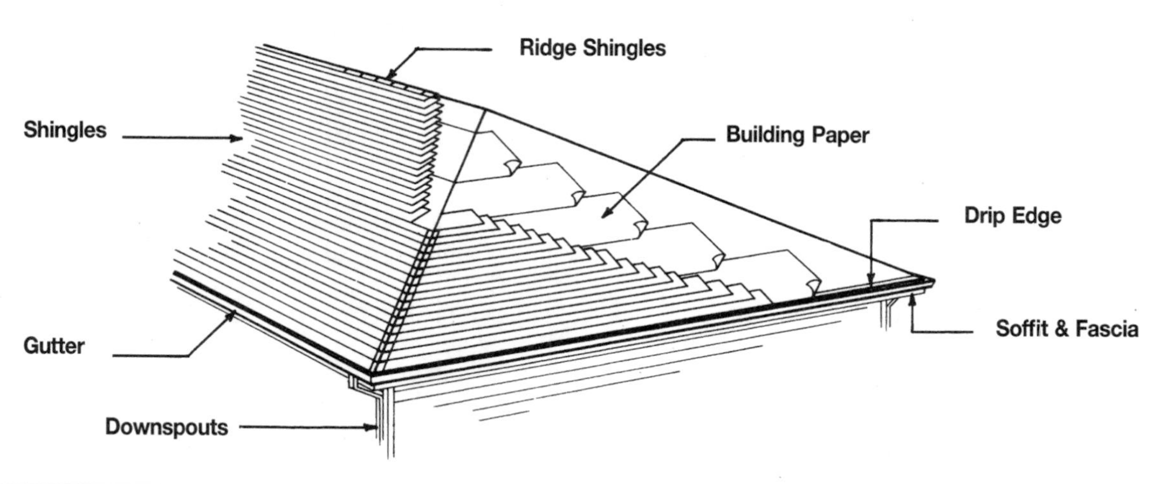

System Description	QUAN.	UNIT	LABOR-HOURS	COST PER S.F. MAT.	COST PER S.F. INST.	COST PER S.F. TOTAL	____S.F. EXTENSIONS
ASPHALT, ROOF SHINGLES, CLASS A							
Shingles, inorganic, class A, 210-235 lb./sq. 4/12 pitch	1.570	S.F.	.023	.48	.77	1.25	
Drip edge, metal, 5″ wide	.122	L.F.	.002	.03	.09	.12	
Building paper, #15 asphalt felt	1.800	S.F.	.002	.04	.08	.12	
Ridge shingles, asphalt	.075	L.F.	.002	.06	.06	.12	
Soffit & fascia, white painted aluminum, 1′ overhang	.120	L.F.	.017	.29	.61	.90	
Gutter, seamless, aluminum, painted	.120	L.F.	.008	.14	.30	.44	
Downspouts, aluminum, painted	.035	L.F.	.002	.04	.06	.10	
TOTAL			.056	1.08	1.97	3.05	
ASPHALT, ROOF SHINGLES, LAMINATED, MULTI-LAYERED							
Laminated, multi-layered, 240-260 lb./sq., 4/12 pitch	1.570	S.F.	.028	.68	.94	1.62	
Drip edge, metal, 5″ wide	.122	L.F.	.002	.03	.09	.12	
Building paper, #15 asphalt felt	1.800	S.F.	.002	.04	.08	.12	
Ridge shingles, asphalt	.075	L.F.	.002	.06	.06	.12	
Soffit & fascia, white painted aluminum, 1′ overhang	.120	L.F.	.017	.29	.61	.90	
Gutter, seamless, aluminum, painted	.120	L.F.	.008	.14	.30	.44	
Downspouts, aluminum, painted	.035	L.F.	.002	.04	.06	.10	
TOTAL			.061	1.28	2.14	3.42	
MISSION TILE, RED							
Mission tile, red, 4/12 pitch	1.570	S.F.	.111	11.12	3.71	14.83	
Drip edge, metal, 5″ wide	.122	L.F.	.002	.03	.09	.12	
Building paper, #15 asphalt felt	1.800	S.F.	.002	.04	.08	.12	
Ridge tiles, clay	.075	L.F.	.003	.71	.10	.81	
Soffit & fascia, white painted aluminum, 1′ overhang	.120	L.F.	.017	.29	.61	.90	
Gutter, seamless, aluminum, painted	.120	L.F.	.008	.14	.30	.44	
Downspouts, aluminum, painted	.035	L.F.	.002	.04	.06	.10	
TOTAL			.145	12.37	4.95	17.32	

Hip Roof Roofing Systems

System Description	QUAN.	UNIT	LABOR-HOURS	COST PER S.F. MAT.	COST PER S.F. INST.	COST PER S.F. TOTAL	S.F. EXTENSIONS
WOOD, CEDAR SHINGLES, NO. 1 PERFECTIONS, 18″ LONG							
Shingles, red cedar, No. 1 perfections, 5″ exp., 4/12 pitch	1.570	S.F.	.047	2.82	1.62	4.44	
Drip edge, metal, 5″ wide	.122	L.F.	.002	.03	.09	.12	
Building paper, #15 asphalt felt	1.800	S.F.	.002	.04	.08	.12	
Ridge shingles, wood, cedar	.075	L.F.	.002	.20	.07	.27	
Soffit & fascia, white painted aluminum, 1′ overhang	.120	L.F.	.017	.29	.61	.90	
Gutter, seamless, aluminum, painted	.120	L.F.	.008	.14	.30	.44	
Downspouts, aluminum, painted	.035	L.F.	.002	.04	.06	.10	
TOTAL			.080	3.56	2.83	6.39	
WOOD, FIRE RETARDANT, CEDAR SHINGLES, NO. 1, 18″ LONG							
Shingles, red cedar, fire retardant, 5″ exp., 4/12 pitch	1.570	S.F.	.047	3.32	1.62	4.94	
Drip edge, metal, 5″ wide	.122	L.F.	.002	.03	.09	.12	
Building paper, #15 asphalt felt	1.800	S.F.	.002	.04	.08	.12	
Ridge shingles, wood, cedar	.075	L.F.	.002	.20	.07	.27	
Soffit & fascia, white painted aluminum, 1′ overhang	.120	L.F.	.017	.29	.61	.90	
Gutter, seamless, aluminum, painted	.120	L.F.	.008	.14	.30	.44	
Downspouts, aluminum, painted	.035	L.F.	.002	.04	.06	.10	
TOTAL			.080	4.06	2.83	6.89	
WOOD SHAKES							
Shingles, hand split, 24″ long, 10″ exp., 4/12 pitch	1.510	S.F.	.051	2.43	1.78	4.21	
Drip edge, metal, 5″ wide	.122	L.F.	.002	.03	.09	.12	
Building paper, #15 asphalt felt	1.800	S.F.	.002	.04	.08	.12	
Ridge shingles, shakes	.075	L.F.	.002	.20	.07	.27	
Soffit & fascia, white painted aluminum, 1′ overhang	.120	L.F.	.017	.29	.61	.90	
Gutter, seamless, aluminum, painted	.012	L.F.	.008	.14	.30	.44	
Downspouts, aluminum, painted	.035	L.F.	.002	.04	.06	.10	
TOTAL			.084	3.17	2.99	6.16	

The prices in these systems are based on a square foot of plan area.
All quantities have been adjusted accordingly.

Hip Roof Roofing System Components

Component Description	QUAN.	UNIT	LABOR-HOURS	COST PER S.F. MAT.	COST PER S.F. INST.	COST PER S.F. TOTAL
Shingles						
Shingles, asphalt, inorganic, class A, 210-235 lb./sq., 4/12 pitch	1.570	S.F.	.023	.48	.77	1.25
8/12 pitch	1.850	S.F.	.028	.57	.91	1.48
Laminated, multi-layered, 240-260 lb./sq., 4/12 pitch	1.570	S.F.	.028	.68	.94	1.62
8/12 pitch	1.850	S.F.	.034	.81	1.11	1.92
Prem. laminated, multi-layered, 260-300 lb./sq., 4/12 pitch	1.570	S.F.	.037	.86	1.21	2.07
8/12 pitch	1.850	S.F.	.043	1.02	1.43	2.45
Clay tile, Spanish tile, red, 4/12 pitch	1.570	S.F.	.071	5.30	2.37	7.67
8/12 pitch	1.850	S.F.	.084	6.25	2.81	9.06
Slate, Buckingham, Virginia, black, 4/12 pitch	1.570	S.F.	.073	9.50	2.43	11.93
8/12 pitch	1.850	S.F.	.087	11.30	2.89	14.19
Vermont, black or grey, 4/12 pitch	1.570	S.F.	.073	6.10	2.43	8.53
8/12 pitch	1.850	S.F.	.087	7.20	2.89	10.09
Wood, red cedar, No.1 5X, 16″ long, 5″ exposure, 4/12 pitch	1.570	S.F.	.051	2.66	1.78	4.44
8/12 pitch	1.850	S.F.	.061	3.15	2.11	5.26
Fire retardant, 4/12 pitch	1.570	S.F.	.051	3.19	1.78	4.97
8/12 pitch	1.850	S.F.	.061	3.78	2.11	5.89

Gambrel Roofing Systems

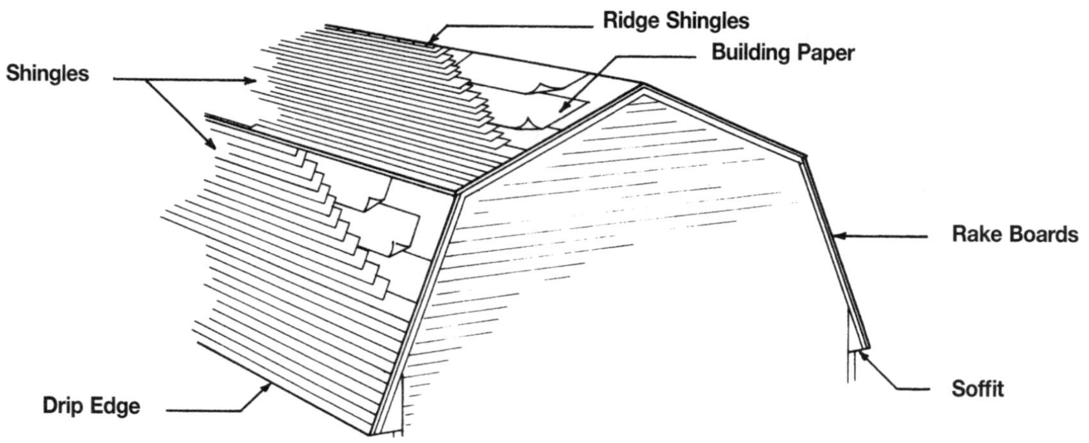

Ridge Shingles
Building Paper
Shingles
Rake Boards
Soffit
Drip Edge

System Description	QUAN.	UNIT	LABOR-HOURS	COST PER S.F. MAT.	COST PER S.F. INST.	COST PER S.F. TOTAL	____S.F. EXTENSIONS
ASPHALT, ROOF SHINGLES, CLASS A							
Shingles, asphalt, inorganic, class A, 210-235 lb./sq.	1.450	S.F.	.022	.45	.72	1.17	
Drip edge, metal, 5″ wide	.146	L.F.	.003	.04	.10	.14	
Building paper, #15 asphalt felt	1.500	S.F.	.002	.04	.07	.11	
Ridge shingles, asphalt	.042	L.F.	.001	.03	.03	.06	
Soffit & fascia, painted aluminum, 1′ overhang	.083	L.F.	.012	.20	.42	.62	
Rake, trim, painted, 1″ x 6″	.063	L.F.	.006	.07	.18	.25	
Gutter, seamless, alumunum, painted	.083	L.F.	.006	.10	.21	.31	
Downspouts, aluminum, painted	.042	L.F.	.002	.05	.07	.12	
TOTAL			.054	.98	1.80	2.78	
ASPHALT, ROOF SHINGLES, LAMINATED, MULTI-LAYERED							
Laminated, multi-layered, 240-260 lb./sq.	1.450	S.F.	.027	.64	.88	1.52	
Drip edge, metal, 5″ wide	.146	L.F.	.003	.04	.10	.14	
Building paper, #15 asphalt felt	1.500	S.F.	.002	.04	.07	.11	
Ridge shingles, asphalt	.042	L.F.	.001	.03	.03	.06	
Soffit & fascia, painted aluminum, 1′ overhang	.083	L.F.	.012	.20	.42	.62	
Rake, trim, painted, 1″ x 6″	.063	L.F.	.006	.07	.18	.25	
Gutter, seamless, alumunum, painted	.083	L.F.	.006	.10	.21	.31	
Downspouts, aluminum, painted	.042	L.F.	.002	.05	.07	.12	
TOTAL			.059	1.17	1.96	3.13	
ASPHALT, ROOF SHINGLES, PREMIUM LAMINATED							
Premium laminated, multi-layered, 260-300 lb./sq.	1.450	S.F.	.034	.80	1.13	1.93	
Drip edge, metal, 5″ wide	.146	L.F.	.003	.04	.10	.14	
Building paper, #15 asphalt felt	1.500	S.F.	.002	.04	.07	.11	
Ridge shingles, asphalt	.042	L.F.	.001	.03	.03	.06	
Soffit & fascia, painted aluminum, 1′ overhang	.083	L.F.	.012	.20	.42	.62	
Rake, trim, painted, 1″ x 6″	.063	L.F.	.006	.07	.18	.25	
Gutter, seamless, alumunum, painted	.083	L.F.	.006	.10	.21	.31	
Downspouts, aluminum, painted	.042	L.F.	.002	.05	.07	.12	
TOTAL			.066	1.33	2.21	3.54	

Gambrel Roofing Systems

System Description	QUAN.	UNIT	LABOR-HOURS	COST PER S.F.			S.F. EXTENSIONS
				MAT.	INST.	TOTAL	
VIRGINIA SLATE							
Slate, Buckingham, Virginia, black	1.450	S.F.	.069	8.93	2.28	11.21	
Drip edge, metal, 5″ wide	.146	L.F.	.003	.04	.10	.14	
Building paper, #15 asphalt felt	1.500	S.F.	.002	.04	.07	.11	
Ridge slate	.042	L.F.	.002	.39	.06	.45	
Soffit & fascia, painted aluminum, 1′ overhang	.083	L.F.	.012	.20	.42	.62	
Rake, trim, painted, 1″ x 6″	.063	L.F.	.006	.07	.18	.25	
Gutter, seamless, alumunum, painted	.083	L.F.	.006	.10	.21	.31	
Downspouts, aluminum, painted	.042	L.F.	.002	.05	.07	.12	
TOTAL			.102	9.82	3.39	13.21	
WOOD, CEDAR SHINGLES, NO. 1 PERFECTIONS, 18″ LONG							
Shingles, wood, red cedar, No. 1 perfections, 5″ exposure	1.450	S.F.	.044	2.64	1.52	4.16	
Drip edge, metal, 5″ wide	.146	L.F.	.003	.04	.10	.14	
Building paper, #15 asphalt felt	1.500	S.F.	.002	.04	.07	.11	
Ridge shingles, wood	.042	L.F.	.001	.11	.04	.15	
Soffit & fascia, white painted aluminum, 1′ overhang	.083	L.F.	.012	.20	.42	.62	
Rake, trim, painted, 1″ x 6″	.063	L.F.	.004	.06	.13	.19	
Gutter, seamless, aluminum, painted	.083	L.F.	.006	.10	.21	.31	
Downspouts, aluminum, painted	.042	L.F.	.002	.05	.07	.12	
TOTAL			.074	3.24	2.56	5.80	
WOOD SHAKES							
Shingles, hand split, 24″ long, 10″ exp., plain	1.450	S.F.	.048	2.28	1.67	3.95	
Drip edge, metal, 5″ wide	.146	L.F.	.003	.04	.10	.14	
Building paper, #15 felt	1.500	S.F.	.002	.04	.07	.11	
Ridge shingles, wood, shakes	.042	S.F.	.001	.11	.04	.15	
Soffit & fascia, white painted aluminum, 1′ overhang	.083	L.F.	.012	.20	.42	.62	
Rake, trim, painted, 1″ x 6″	.063	L.F.	.004	.06	.13	.19	
Gutter, seamless, aluminum, painted	.083	L.F.	.006	.10	.21	.31	
Downspouts, aluminum, painted	.042	L.F.	.002	.05	.07	.12	
TOTAL			.078	2.88	2.71	5.59	

The prices in this system are based on a square foot of plan area. All quantities have been adjusted accordingly.

Gambrel Roofing System Components

Component Description	QUAN.	UNIT	LABOR-HOURS	COST PER S.F.		
				MAT.	INST.	TOTAL
Shingles						
Wood, red cedar, No.1 5X, 16″ long, 5″ exposure, plain	1.450	S.F.	.048	2.49	1.67	4.16
Fire retardant	1.450	S.F.	.048	2.99	1.67	4.66
Shakes, hand split, 24″ long, 10″ exposure, plain	1.450	S.F.	.048	2.28	1.67	3.95
18″ long, 8″ exposure, plain	1.450	S.F.	.060	1.61	2.09	3.70
Soffit & fascia						
Soffit & fascia, aluminum, vented, 1′ overhang	.083	L.F.	.012	.20	.42	.62
2′ overhang	.083	L.F.	.013	.29	.46	.75
Wood board fascia, plywood soffit, 1′ overhang	.083	L.F.	.004	.02	.11	.13
2′ overhang	.083	L.F.	.006	.03	.17	.20

Mansard Roofing Systems

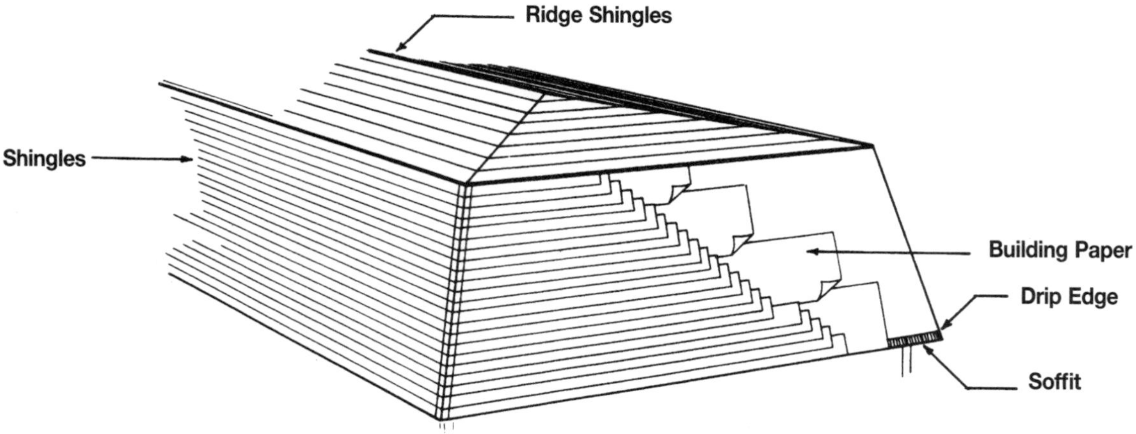

Ridge Shingles
Shingles
Building Paper
Drip Edge
Soffit

System Description	QUAN.	UNIT	LABOR-HOURS	COST PER S.F. MAT.	INST.	TOTAL	____S.F. EXTENSIONS
ASPHALT, ROOF SHINGLES, CLASS A							
Shingles, standard inorganic class A 210-235 lb./sq.	2.210	S.F.	.032	.66	1.06	1.72	
Drip edge, metal, 5" wide	.122	L.F.	.002	.03	.09	.12	
Building paper, #15 asphalt felt	2.300	S.F.	.003	.05	.10	.15	
Ridge shingles, asphalt	.090	L.F.	.002	.07	.07	.14	
Soffit & fascia, white painted aluminum, 1' overhang	.122	L.F.	.018	.29	.62	.91	
Gutter, seamless, aluminum, painted	.122	L.F.	.008	.15	.30	.45	
Downspouts, aluminum, painted	.042	L.F.	.002	.05	.07	.12	
TOTAL			.067	1.30	2.31	3.61	
ASPHALT, ROOF SHINGLES, PREMIUM LAMINATED							
Premium laminated, multi-layered, 260-300 lb./sq.	2.210	S.F.	.050	1.18	1.66	2.84	
Drip edge, metal, 5" wide	.122	L.F.	.002	.03	.09	.12	
Building paper, #15 asphalt felt	2.300	S.F.	.003	.05	.10	.15	
Ridge shingles, asphalt	.090	L.F.	.002	.07	.07	.14	
Soffit & fascia, white painted aluminum, 1' overhang	.122	L.F.	.018	.29	.62	.91	
Gutter, seamless, aluminum, painted	.122	L.F.	.008	.15	.30	.45	
Downspouts, aluminum, painted	.042	L.F.	.002	.05	.07	.12	
TOTAL			.085	1.82	2.91	4.73	
VERMONT SLATE, BLACK OR GREY							
Vermont, black or grey	2.210	S.F.	.101	8.36	3.34	11.70	
Drip edge, metal, 5" wide	.122	L.F.	.002	.03	.09	.12	
Building paper, #15 asphalt felt	2.300	S.F.	.003	.05	.10	.15	
Ridge slate	.090	L.F.	.004	.84	.12	.96	
Soffit & fascia, white painted aluminum, 1' overhang	.122	L.F.	.018	.29	.62	.91	
Gutter, seamless, aluminum, painted	.122	L.F.	.008	.15	.30	.45	
Downspouts, aluminum, painted	.042	L.F.	.002	.05	.07	.12	
TOTAL			.138	9.77	4.64	14.41	

Mansard Roofing Systems

System Description	QUAN.	UNIT	LABOR-HOURS	COST PER S.F. MAT.	COST PER S.F. INST.	COST PER S.F. TOTAL	____S.F. EXTENSIONS
WOOD, CEDAR SHINGLES, NO. 1 PERFECTIONS, 18″ LONG							
Shingles, wood, red cedar, No. 1 perfections, 5″ exposure	2.210	S.F.	.064	3.87	2.22	6.09	
Drip edge, metal, 5″ wide	.122	L.F.	.002	.03	.09	.12	
Building paper, #15 asphalt felt	2.300	S.F.	.003	.05	.10	.15	
Ridge shingles, wood	.090	L.F.	.003	.24	.09	.33	
Soffit & fascia, white painted aluminum, 1′ overhang	.122	L.F.	.018	.29	.62	.91	
Gutter, seamless, aluminum, painted	.122	L.F.	.008	.15	.30	.45	
Downspouts, aluminum, painted	.042	L.F.	.002	.05	.07	.12	
TOTAL			.100	4.68	3.49	8.17	
WOOD, CEDAR SHINGLES, FIRE RETARD, NO. 1 PERFECTIONS							
Shingles, wood, red cedar, fire retardant, 5″ exp.	2.210	S.F.	.070	4.38	2.44	6.82	
Drip edge, metal, 5″ wide	.122	L.F.	.002	.03	.09	.12	
Building paper, #15 asphalt felt	2.300	S.F.	.003	.05	.10	.15	
Ridge shingles, wood	.090	L.F.	.003	.24	.09	.33	
Soffit & fascia, white painted aluminum, 1′ overhang	.122	L.F.	.018	.29	.62	.91	
Gutter, seamless, aluminum, painted	.122	L.F.	.008	.15	.30	.45	
Downspouts, aluminum, painted	.042	L.F.	.002	.05	.07	.12	
TOTAL			.106	5.19	3.71	8.90	
WOOD SHAKES							
Shingles, hand split, 24″ long, 10″ exposure, plain	2.210	S.F.	.070	3.34	2.44	5.78	
Drip edge, metal, 5″ wide	.122	L.F.	.002	.03	.09	.12	
Building paper, #15 asphalt felt	2.300	S.F.	.003	.05	.10	.15	
Ridge shingles, shakes	.090	L.F.	.003	.24	.09	.33	
Soffit & fascia, white painted aluminum, 1′ overhang	.122	L.F.	.018	.29	.62	.91	
Gutter, seamless, aluminum, painted	.122	L.F.	.008	.15	.30	.45	
Downspouts, aluminum, painted	.042	L.F.	.002	.05	.07	.12	
TOTAL			.106	4.15	3.71	7.86	

The prices in these systems are based on a square foot of plan area.
All quantities have been adjusted accordingly.

Mansard Roofing System Components

Component Description	QUAN.	UNIT	LABOR-HOURS	COST PER S.F. MAT.	COST PER S.F. INST.	COST PER S.F. TOTAL
Shingles						
Laminated, multi-layered, 240-260 lb./sq.	2.210	S.F.	.039	.94	1.29	2.23
Slate Buckingham, Virginia, black	2.210	S.F.	.101	13.10	3.34	16.44
Wood, red cedar, No.1 5X, 16″ long, 5″ exposure, plain	2.210	S.F.	.070	3.65	2.44	6.09
Fire retardant	2.210	S.F.	.070	4.38	2.44	6.82
Resquared & rebutted, 18″ long, 6″ exposure, plain	2.210	S.F.	.059	4.82	2.05	6.87
Fire retardant	2.210	S.F.	.059	5.50	2.05	7.55
Shakes, hand split, 18″ long, 8″ exposure, plain	2.210	S.F.	.088	2.35	3.06	5.41
Fire retardant	2.210	S.F.	.088	3.08	3.06	6.14
Downspout						
Downspout 2″ x 3″, aluminum, one story house	.042	L.F.	.002	.04	.07	.11
Two story house	.070	L.F.	.003	.07	.11	.18
Vinyl, one story house	.042	L.F.	.002	.04	.07	.11
Two story house	.070	L.F.	.003	.07	.11	.18

Shed Roofing Systems

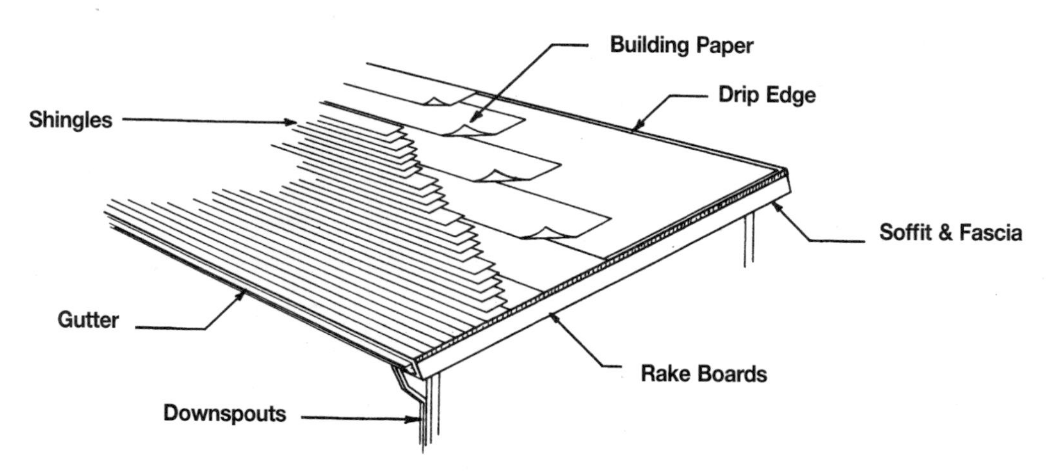

System Description	QUAN.	UNIT	LABOR-HOURS	COST PER S.F. MAT.	COST PER S.F. INST.	COST PER S.F. TOTAL	____S.F. EXTENSIONS
ASPHALT, ROOF SHINGLES, CLASS A							
Shingles, inorganic class A 210-235 lb./sq. 4/12 pitch	1.230	S.F.	.019	.39	.62	1.01	
Drip edge, metal, 5″ wide	.100	L.F.	.002	.02	.07	.09	
Building paper, #15 asphalt felt	1.300	S.F.	.002	.03	.06	.09	
Soffit & fascia, white painted aluminum, 1′ overhang	.080	L.F.	.012	.19	.40	.59	
Rake trim, painted, 1″ x 6″	.043	L.F.	.004	.04	.12	.16	
Gutter, seamless, aluminum, painted	.040	L.F.	.003	.05	.10	.15	
Downspouts, painted aluminum	.020	L.F.	.001	.02	.03	.05	
TOTAL			.043	.74	1.40	2.14	
ASPHALT, ROOF SHINGLES, PREMIUM LAMINATED							
Shingles, premium laminated, 210-235 lb./sq. 4/12 pitch	1.230	S.F.	.027	.64	.91	1.55	
Drip edge, metal, 5″ wide	.100	L.F.	.002	.02	.07	.09	
Building paper, #15 asphalt felt	1.300	S.F.	.002	.03	.06	.09	
Soffit & fascia, white painted aluminum, 1′ overhang	.080	L.F.	.012	.19	.40	.59	
Rake trim, painted, 1″ x 6″	.043	L.F.	.004	.04	.12	.16	
Gutter, seamless, aluminum, painted	.040	L.F.	.003	.05	.10	.15	
Downspouts, painted aluminum	.020	L.F.	.001	.02	.03	.05	
TOTAL			.051	.99	1.69	2.68	
CLAY TILE, SPANISH TILE							
Clay tile, Spanish tile, red, 4/12 pitch	1.230	S.F.	.053	3.96	1.78	5.74	
Drip edge, metal, 5″ wide	.100	L.F.	.002	.02	.07	.09	
Building paper, #15 asphalt felt	1.300	S.F.	.002	.03	.06	.09	
Soffit & fascia, white painted aluminum, 1′ overhang	.080	L.F.	.012	.19	.40	.59	
Rake trim, painted, 1″ x 6″	.043	L.F.	.004	.04	.12	.16	
Gutter, seamless, aluminum, painted	.040	L.F.	.003	.05	.10	.15	
Downspouts, painted aluminum	.020	L.F.	.001	.02	.03	.05	
TOTAL			.077	4.31	2.56	6.87	

Shed Roofing Systems

System Description	QUAN.	UNIT	LABOR-HOURS	COST PER S.F.			____S.F. EXTENSIONS
				MAT.	INST.	TOTAL	
WOOD, CEDAR SHINGLES, NO. 1 PERFECTIONS, 18″ LONG							
Shingles, red cedar, No. 1 perfections, 5″ exp., 4/12 pitch	1.230	S.F.	.035	2.11	1.21	3.32	
Drip edge, metal, 5″ wide	.100	L.F.	.002	.02	.07	.09	
Building paper, #15 asphalt felt	1.300	S.F.	.002	.03	.06	.09	
Soffit & fascia, white painted aluminum, 1′ overhang	.080	L.F.	.012	.19	.40	.59	
Rake trim, painted, 1″ x 6″	.043	L.F.	.003	.04	.09	.13	
Gutter, seamless, aluminum, painted	.040	L.F.	.003	.05	.10	.15	
Downspouts, painted aluminum	.020	L.F.	.001	.02	.03	.05	
TOTAL			.058	2.46	1.96	4.42	
WOOD, CEDAR SHINGLES, RESQUARED & REBUTTED 18″ LONG							
Resquared & rebutted, 18″ long, 5″ exposure, 4/12 pitch	1.230	S.F.	.032	2.63	1.12	3.75	
Drip edge, metal, 5″ wide	.100	L.F.	.002	.02	.07	.09	
Building paper, #15 asphalt felt	1.300	S.F.	.002	.03	.06	.09	
Soffit & fascia, white painted aluminum, 1′ overhang	.080	L.F.	.012	.19	.40	.59	
Rake trim, painted, 1″ x 6″	.043	L.F.	.003	.04	.09	.13	
Gutter, seamless, aluminum, painted	.040	L.F.	.003	.05	.10	.15	
Downspouts, painted aluminum	.020	L.F.	.001	.02	.03	.05	
TOTAL			.055	2.98	1.87	4.85	
WOOD SHAKES							
Shingles, hand split, 24″ long, 10″ exp., 4/12 pitch	1.230	S.F.	.038	1.82	1.33	3.15	
Drip edge, metal, 5″ wide	.100	L.F.	.002	.02	.07	.09	
Building paper, #15 asphalt felt	1.300	S.F.	.002	.03	.06	.09	
Soffit & fascia, white painted aluminum, 1′ overhang	.080	L.F.	.012	.19	.40	.59	
Rake trim, painted, 1″ x 6″	.043	L.F.	.003	.04	.09	.13	
Gutter, seamless, aluminum, painted	.040	L.F.	.003	.05	.10	.15	
Downspouts, painted aluminum	.020	L.F.	.001	.02	.03	.05	
TOTAL			.061	2.17	2.08	4.25	

The prices in these systems are based on a square foot of plan area.
All quantities have been adjusted accordingly.

Shed Roofing System Components

Component Description	QUAN.	UNIT	LABOR-HOURS	COST PER S.F.		
				MAT.	INST.	TOTAL
Shingles						
Shingles, asphalt, inorganic, class A, 210-235 lb./sq., 4/12 pitch	1.230	S.F.	.017	.36	.58	.94
8/12 pitch	1.330	S.F.	.019	.39	.62	1.01
Laminated, multi-layered, 240-260 lb./sq. 4/12 pitch	1.230	S.F.	.021	.51	.70	1.21
8/12 pitch	1.330	S.F.	.023	.55	.76	1.31
Premium laminated, multi-layered, 260-300 lb./sq. 4/12 pitch	1.230	S.F.	.027	.64	.91	1.55
8/12 pitch	1.330	S.F.	.030	.70	.98	1.68
Mission tile, red, 4/12 pitch	1.230	S.F.	.083	8.35	2.78	11.13
8/12 pitch	1.330	S.F.	.090	9.05	3.02	12.07
French tile, red, 4/12 pitch	1.230	S.F.	.071	7.55	2.38	9.93
8/12 pitch	1.330	S.F.	.077	8.20	2.57	10.77
Slate, Buckingham, Virginia, black, 4/12 pitch	1.230	S.F.	.055	7.15	1.82	8.97
8/12 pitch	1.330	S.F.	.059	7.75	1.98	9.73
Vermont, black or grey, 4/12 pitch	1.230	S.F.	.055	4.56	1.82	6.38
8/12 pitch	1.330	S.F.	.059	4.94	1.98	6.92
Wood, red cedar, No.1 5X, 16″ long, 5″ exposure, 4/12 pitch	1.230	S.F.	.038	1.99	1.33	3.32
8/12 pitch	1.330	S.F.	.042	2.16	1.44	3.60

Gable Dormer Roofing Systems

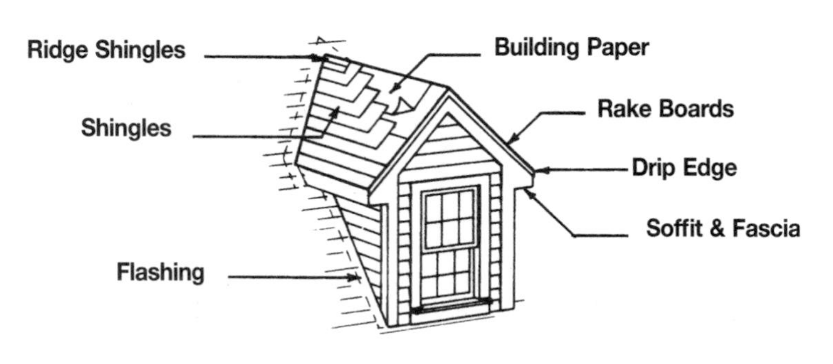

System Description	QUAN.	UNIT	LABOR-HOURS	COST PER S.F. MAT.	COST PER S.F. INST.	COST PER S.F. TOTAL	____S.F. EXTENSIONS
ASPHALT, ROOF SHINGLES, CLASS A							
Shingles, standard inorganic class A 210-235 lb./sq	1.400	S.F.	.020	.42	.67	1.09	
Drip edge, metal, 5'' wide	.220	L.F.	.004	.05	.15	.20	
Building paper, #15 asphalt felt	1.500	S.F.	.002	.04	.07	.11	
Ridge shingles, asphalt	.280	L.F.	.007	.22	.22	.44	
Soffit & fascia, aluminum, vented	.220	L.F.	.032	.52	1.11	1.63	
Flashing, aluminum, mill finish, .013'' thick	1.500	S.F.	.083	.59	3.09	3.68	
TOTAL			.148	1.84	5.31	7.15	
ASPHALT, ROOF SHINGLES, PREMIUM LAMINATED							
Premium laminated, multi-layered, 260-300 lb./sq.	1.400	S.F.	.032	.75	1.06	1.81	
Drip edge, metal, 5'' wide	.220	L.F.	.004	.05	.15	.20	
Building paper, #15 asphalt felt	1.500	S.F.	.002	.04	.07	.11	
Ridge shingles, asphalt	.280	L.F.	.007	.22	.22	.44	
Soffit & fascia, aluminum, vented	.220	L.F.	.032	.52	1.11	1.63	
Flashing, aluminum, mill finish, .013'' thick	1.500	S.F.	.083	.59	3.09	3.68	
TOTAL			.160	2.17	5.70	7.87	
WOOD, CEDAR, NO. 1 PERFECTIONS							
Shingles, red cedar, No.1 perfections, 18'' long, 5'' exp.	1.400	S.F.	.041	2.46	1.41	3.87	
Drip edge, metal, 5'' wide	.220	L.F.	.004	.05	.15	.20	
Building paper, #15 asphalt felt	1.500	S.F.	.002	.04	.07	.11	
Ridge shingles, wood	.280	L.F.	.008	.74	.28	1.02	
Soffit & fascia, aluminum, vented	.220	L.F.	.032	.52	1.11	1.63	
Flashing, aluminum, mill finish, .013'' thick	1.500	S.F.	.083	.59	3.09	3.68	
TOTAL			.170	4.40	6.11	10.51	
SLATE, BUCKINGHAM, BLACK							
Shingles, Buckingham, Virginia, black	1.400	S.F.	.064	8.33	2.13	10.46	
Drip edge, metal, 5'' wide	.220	L.F.	.004	.05	.15	.20	
Building paper, #15 asphalt felt	1.500	S.F.	.002	.04	.07	.11	
Ridge shingles, slate	.280	L.F.	.011	2.62	.37	2.99	
Soffit & fascia, aluminum, vented	.220	L.F.	.032	.52	1.11	1.63	
Flashing, copper, 16 oz.	1.500	S.F.	.104	3.60	3.90	7.50	
TOTAL			.217	15.16	7.73	22.89	

Gable Dormer Roofing Systems

System Description	QUAN.	UNIT	LABOR-HOURS	COST PER S.F.			____S.F. EXTENSIONS
				MAT.	INST.	TOTAL	
WOOD, CEDAR, RESQUARED AND REBUTTED							
Resquared & rebutted, 18" long, 5" exposure	1.400	S.F.	.037	3.07	1.30	4.37	
Drip edge, metal, 5" wide	.220	L.F.	.004	.05	.15	.20	
Building paper, #15 asphalt felt	1.500	S.F.	.002	.04	.07	.11	
Ridge shingles, wood	.280	L.F.	.011	2.62	.37	2.99	
Soffit & fascia, aluminum, vented	.220	L.F.	.032	.52	1.11	1.63	
Flashing, copper, 16 oz.	1.500	S.F.	.104	3.60	3.90	7.50	
TOTAL			.190	9.90	6.90	16.80	
WOOD SHAKES							
Shingles, wood shakes hand split, 24" long, 10" exp.	1.400	S.F.	.045	2.13	1.55	3.68	
Drip edge, metal, 5" wide	.220	L.F.	.004	.05	.15	.20	
Building paper, #15 asphalt felt	1.500	S.F.	.002	.04	.07	.11	
Ridge shingles, shakes	.280	L.F.	.011	2.62	.37	2.99	
Soffit & fascia, aluminum, vented	.220	L.F.	.032	.52	1.11	1.63	
Flashing, copper, 16 oz.	1.500	S.F.	.104	3.60	3.90	7.50	
TOTAL			.198	8.96	7.15	16.11	

The prices in these systems are based on a square foot of plan area under the dormer roof.

Gable Dormer Roofing System Components

Component Description	QUAN.	UNIT	LABOR-HOURS	COST PER S.F.		
				MAT.	INST.	TOTAL
Shingles						
Shingles, asphalt, standard, inorganic, class A, 210-235 lb./sq.	1.400	S.F.	.020	.42	.67	1.09
Laminated, multi-layered, 240-260 lb./sq.	1.400	S.F.	.025	.60	.82	1.42
Clay tile, Spanish tile, red	1.400	S.F.	.062	4.62	2.07	6.69
Mission tile, red	1.400	S.F.	.097	9.75	3.25	13.00
French tile, red	1.400	S.F.	.083	8.80	2.77	11.57
Slate Buckingham, Virginia, black	1.400	S.F.	.064	8.35	2.13	10.48
Vermont, black or grey	1.400	S.F.	.064	5.30	2.13	7.43
Wood, red cedar, No.1 5X, 16" long, 5" exposure	1.400	S.F.	.045	2.32	1.55	3.87
Fire retardant	1.400	S.F.	.045	2.78	1.55	4.33
Shakes hand split, 24" long, 10" exposure	1.400	S.F.	.045	2.13	1.55	3.68
18" long, 8" exposure	1.400	S.F.	.056	1.50	1.95	3.45
Flashing						
Flashing, aluminum, .013" thick	1.500	S.F.	.083	.59	3.09	3.68
.032" thick	1.500	S.F.	.083	1.71	3.09	4.80
.040" thick	1.500	S.F.	.083	2.40	3.09	5.49
.050" thick	1.500	S.F.	.083	3.03	3.09	6.12
Copper, 16 oz.	1.500	S.F.	.104	3.60	3.90	7.50
20 oz.	1.500	S.F.	.109	6.45	4.07	10.52
24 oz.	1.500	S.F.	.114	7.75	4.26	12.01
32 oz.	1.500	S.F.	.120	10.30	4.47	14.77

Shed Dormer Roofing Systems

System Description	QUAN.	UNIT	LABOR-HOURS	COST PER S.F. MAT.	COST PER S.F. INST.	COST PER S.F. TOTAL	____S.F. EXTENSIONS
ASPHALT, ROOF SHINGLES, CLASS A							
Shingles, standard inorganic class A 210-235 lb./sq.	1.100	S.F.	.016	.33	.53	.86	
Drip edge, aluminum, 5″ wide	.250	L.F.	.005	.06	.18	.24	
Building paper, #15 asphalt felt	1.200	S.F.	.002	.03	.05	.08	
Soffit & fascia, aluminum, vented, 1′ overhang	.250	L.F.	.036	.60	1.26	1.86	
Flashing, aluminum, mill finish, 0.013″ thick	.800	L.F.	.044	.31	1.65	1.96	
TOTAL			.103	1.33	3.67	5.00	
ASPHALT, ROOF SHINGLES, PREMIUM LAMINATED							
Premium laminated, multi-layered, 260-300 lb./sq.	1.100	S.F.	.025	.59	.83	1.42	
Drip edge, aluminum, 5″ wide	.250	L.F.	.005	.06	.18	.24	
Building paper, #15 asphalt felt	1.200	S.F.	.002	.03	.05	.08	
Soffit & fascia, aluminum, vented, 1′ overhang	.250	L.F.	.036	.60	1.26	1.86	
Flashing, aluminum, mill finish, 0.013″ thick	.800	L.F.	.044	.31	1.65	1.96	
TOTAL			.112	1.59	3.97	5.56	
CLAY TILE, SPANISH TILE							
Clay tile, Spanish tile, red	1.100	S.F.	.049	3.63	1.63	5.26	
Drip edge, aluminum, 5″ wide	.250	L.F.	.005	.06	.18	.24	
Building paper, #15 asphalt felt	1.200	S.F.	.002	.03	.05	.08	
Soffit & fascia, aluminum, vented, 1′ overhang	.250	L.F.	.036	.60	1.26	1.86	
Flashing, copper, 16 oz.	.800	L.F.	.056	1.92	2.08	4.00	
TOTAL			.148	6.24	5.20	11.44	
WOOD, CEDAR, NO. 1 PERFECTIONS, 18″ LONG							
Shingles, wood, red cedar, #1 perfections, 5″ exposure	1.100	S.F.	.032	1.94	1.11	3.05	
Drip edge, aluminum, 5″ wide	.250	L.F.	.005	.06	.18	.24	
Building paper, #15 asphalt felt	1.200	S.F.	.002	.03	.05	.08	
Soffit & fascia, aluminum, vented, 1′ overhang	.250	L.F.	.036	.60	1.26	1.86	
Flashing, aluminum, mill finish, 0.013″ thick	.800	L.F.	.044	.31	1.65	1.96	
TOTAL			.119	2.94	4.25	7.19	

Shed Dormer Roofing Systems

System Description	QUAN.	UNIT	LABOR-HOURS	COST PER S.F. MAT.	INST.	TOTAL	S.F. EXTENSIONS
WOOD, CEDAR, RESQUARED AND REBUTTED, 18″ LONG							
Resquared & rebutted, 18″ long, 5″ exposure	1.100	S.F.	.029	2.41	1.02	3.43	
Drip edge, aluminum, 5″ wide	.250	L.F.	.005	.06	.18	.24	
Building paper, #15 asphalt felt	1.200	S.F.	.002	.03	.05	.08	
Soffit & fascia, aluminum, vented, 1′ overhang	.250	L.F.	.036	.60	1.26	1.86	
Flashing, aluminum, mill finish, 0.013″ thick	.800	L.F.	.044	.31	1.65	1.96	
TOTAL			.116	3.41	4.16	7.57	
WOOD SHAKES							
Shingles, wood shakes hand split, 24″ long, 10″ exp.	1.100	S.F.	.035	1.67	1.22	2.89	
Drip edge, metal 5″ wide	.250	L.F.	.005	.06	.18	.24	
Building paper, #15 asphalt felt	1.200	S.F.	.002	.03	.05	.08	
Soffit & fascia, aluminum, vented, 1′ overhang	.250	L.F.	.036	.60	1.26	1.86	
Flashing, aluminum, mill finish, 0.013″ thick	.800	L.F.	.044	.31	1.65	1.96	
TOTAL			.122	2.67	4.36	7.03	

The prices in this system are based on a square foot of plan area under the dormer roof.

Shed Dormer Roofing System Components

Component Description	QUAN.	UNIT	LABOR-HOURS	COST PER S.F. MAT.	INST.	TOTAL
Shingles						
Shingles, asphalt, standard, inorganic, class A, 210-235 lb./sq.	1.100	S.F.	.016	.33	.53	.86
Laminated, multi-layered, 240-260 lb./sq.	1.100	S.F.	.020	.47	.64	1.11
Premium laminated, multi-layered, 260-300 lb./sq.	1.100	S.F.	.025	.59	.83	1.42
Clay tile, Spanish tile, red	1.100	S.F.	.049	3.63	1.63	5.26
Mission tile, red	1.100	S.F.	.077	7.65	2.55	10.20
French tile, red	1.100	S.F.	.065	6.95	2.18	9.13
Slate Buckingham, Virginia, black	1.100	S.F.	.050	6.55	1.67	8.22
Vermont, black or grey	1.100	S.F.	.050	4.18	1.67	5.85
Wood, red cedar, No. 1 5X, 16″ long, 5″ exposure	1.100	S.F.	.035	1.83	1.22	3.05
Fire retardant	1.100	S.F.	.035	2.19	1.22	3.41
Shakes hand split, 24″ long, 10″ exposure	1.100	S.F.	.035	1.67	1.22	2.89
Fire retardant	1.100	S.F.	.035	2.03	1.22	3.25
Flashing						
Flashing, aluminum, .013″ thick	.800	L.F.	.044	.31	1.65	1.96
.032″ thick	.800	L.F.	.044	.91	1.65	2.56
.040″ thick	.800	L.F.	.044	1.28	1.65	2.93
.050″ thick	.800	L.F.	.044	1.62	1.65	3.27
Copper, 16 oz.	.800	L.F.	.056	1.92	2.08	4.00
20 oz.	.800	L.F.	.058	3.43	2.17	5.60
24 oz.	.800	L.F.	.061	4.12	2.27	6.39
32 oz.	.800	L.F.	.064	5.50	2.38	7.88

Skylight/Skywindow Systems

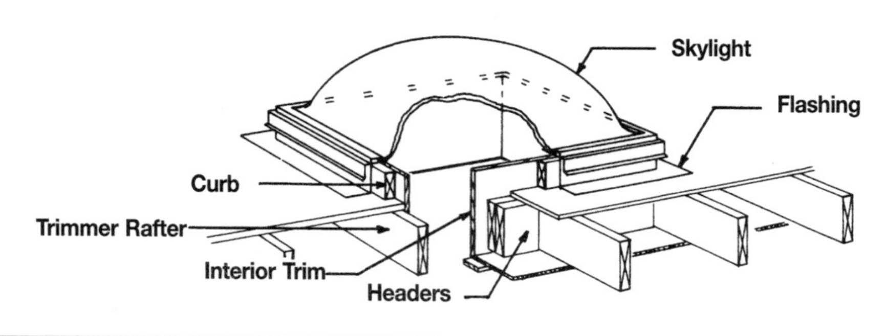

System Description	QUAN.	UNIT	LABOR-HOURS	COST EACH MAT.	COST EACH INST.	COST EACH TOTAL	____EA. EXTENSIONS
SKYLIGHT, FIXED, 32″ X 32″							
Skylight, fixed bubble, insulating, 32″ x 32″	1.000	Ea.	1.422	96.35	44.44	140.79	
Trimmer rafters, 2″ x 6″	28.000	L.F.	.448	18.20	15.68	33.88	
Headers, 2″ x 6″	6.000	L.F.	.267	3.90	9.30	13.20	
Curb, 2″ x 4″	12.000	L.F.	.154	5.28	5.40	10.68	
Flashing, aluminum, .013″ thick	13.500	S.F.	.745	5.27	27.81	33.08	
Trim, interior casing, painted	12.000	L.F.	.696	11.04	23.04	34.08	
TOTAL			3.732	140.04	125.67	265.71	
SKYLIGHT, FIXED, 48″ X 48″							
Skylight, fixed bubble, insulating, 48″ x 48″	1.000	Ea.	1.296	125.60	40.64	166.24	
Trimmer rafters, 2″ x 6″	28.000	L.F.	.448	18.20	15.68	33.88	
Headers, 2″ x 6″	8.000	L.F.	.356	5.20	12.40	17.60	
Curb, 2″ x 4″	16.000	L.F.	.205	7.04	7.20	14.24	
Flashing, aluminum, .013″ thick	16.000	S.F.	.883	6.24	32.96	39.20	
Trim, interior casing, painted	16.000	L.F.	.927	14.72	30.72	45.44	
TOTAL			4.115	177.00	139.60	316.60	
SKYLIGHT, VENTILATING, 36″ X 36″							
Skylight, ventilating bubble insulating, 36″ x 36″	1.000	Ea.	2.667	375.00	83.50	458.50	
Trimmer rafters, 2″ x 6″	28.000	L.F.	.448	18.20	15.68	33.88	
Headers, 2″ x 6″	6.000	L.F.	.356	5.20	12.40	17.60	
Curb, 2″ x 4″	12.000	L.F.	.205	7.04	7.20	14.24	
Flashing, aluminum, .013″ thick	13.500	S.F.	.883	6.24	32.96	39.20	
Trim, interior casing, painted	12.000	L.F.	.927	14.72	30.72	45.44	
TOTAL			5.486	426.40	182.46	608.86	
SKYLIGHT, VENTILATING, 36″ X 52″							
Skylight, ventilating bubble insulating, 36″ x 52″	1.000	Ea.	3.200	475.00	100.00	575.00	
Trimmer rafters, 2″ x 6″	28.000	L.F.	.448	18.20	15.68	33.88	
Headers, 2″ x 6″	8.000	L.F.	.356	5.20	12.40	17.60	
Curb, 2″ x 4″	16.000	L.F.	.205	7.04	7.20	14.24	
Flashing, aluminum, .013″ thick	16.000	S.F.	.883	6.24	32.96	39.20	
Trim, interior casing, painted	16.000	L.F.	.927	14.72	30.72	45.44	
TOTAL			6.019	526.40	198.96	725.36	

Skylight/Skywindow Systems

System Description	QUAN.	UNIT	LABOR-HOURS	COST EACH MAT.	COST EACH INST.	COST EACH TOTAL	____EA. EXTENSIONS
SKYWINDOW, OPERATING, 24″ X 48″							
Skywindow, operating, thermopane glass, 24″ x 48″	1.000	Ea.	3.200	550.00	100.00	650.00	
Trimmer rafters, 2″ x 6″	28.000	L.F.	.448	18.20	15.68	33.88	
Headers, 2″ x 6″	8.000	L.F.	.267	3.90	9.30	13.20	
Curb, 2″ x 4″	14.000	L.F.	.179	6.16	6.30	12.46	
Flashing, aluminum, .013″ thick	14.000	S.F.	.772	5.46	28.84	34.30	
Trim, interior casing, painted	14.000	L.F.	.811	12.88	26.88	39.76	
TOTAL			5.677	596.60	187.00	783.60	
SKYWINDOW, OPERATING, 32″ X 48″							
Skywindow, operating, thermopane glass, 32″ x 48″	1.000	Ea.	3.556	575.00	111.00	686.00	
Trimmer rafters, 2″ x 6″	28.000	L.F.	.448	18.20	15.68	33.88	
Headers, 2″ x 6″	8.000	L.F.	.356	5.20	12.40	17.60	
Curb, 2″ x 4″	14.000	L.F.	.179	6.16	6.30	12.46	
Flashing, aluminum, .013″ thick	14.000	S.F.	.772	5.46	28.84	34.30	
Trim, interior casing, painted	14.000	L.F.	.811	12.88	26.88	39.76	
TOTAL			6.122	622.90	201.10	824.00	

The prices in these systems are on a cost each basis.

Skylight/Skywindow System Components

Component Description	QUAN.	UNIT	LABOR-HOURS	COST EACH MAT.	COST EACH INST.	COST EACH TOTAL
Skylights						
Skylight, fixed bubble insulating, 24″ x 24″	1.000	Ea.	.800	54.00	25.00	79.00
32″ x 48″	1.000	Ea.	.864	83.50	27.00	110.50
Ventilating bubble insulating, 36″ x 36″	1.000	Ea.	2.667	375.00	83.50	458.50
52″ x 52″	1.000	Ea.	2.667	565.00	83.50	648.50
28″ x 52″	1.000	Ea.	3.200	440.00	100.00	540.00
36″ x 52″	1.000	Ea.	3.200	475.00	100.00	575.00
Trimer rafters						
Trimmer rafters, 2″ x 6″	28.000	L.F.	.448	18.20	15.70	33.90
2″ x 8″	28.000	L.F.	.472	25.50	16.50	42.00
2″ x 10″	28.000	L.F.	.711	37.50	24.50	62.00
Headers						
Headers, 24″ window, 2″ x 6″	4.000	L.F.	.178	2.60	6.20	8.80
2″ x 8″	4.000	L.F.	.188	3.64	6.55	10.19
2″ x 10″	4.000	L.F.	.200	5.35	6.95	12.30
32″ window, 2″ x 6″	6.000	L.F.	.267	3.90	9.30	13.20
2″ x 8″	6.000	L.F.	.282	5.45	9.85	15.30
2″ x 10″	6.000	L.F.	.300	8.05	10.45	18.50
48″ window, 2″ x 6″	8.000	L.F.	.356	5.20	12.40	17.60
2″ x 8″	8.000	L.F.	.376	7.30	13.10	20.40
2″ x 10″	8.000	L.F.	.400	10.70	13.90	24.60

Interior Systems

Drywall & Thincoat Wall Systems

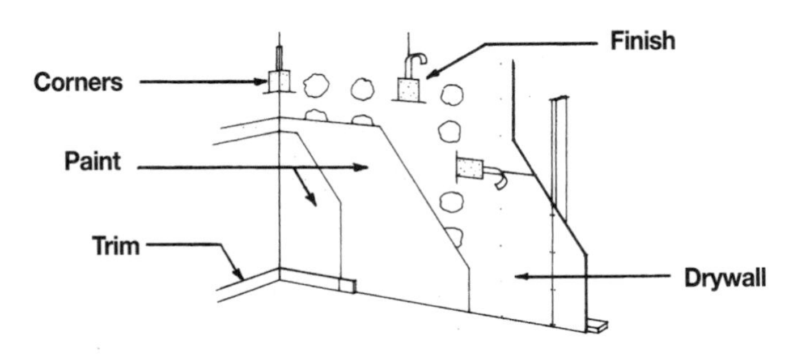

System Description	QUAN.	UNIT	LABOR-HOURS	COST PER S.F.			______S.F. EXTENSIONS
				MAT.	INST.	TOTAL	
1/2″ SHEETROCK, TAPED & FINISHED							
Drywall, 1/2″ thick, standard	1.000	S.F.	.008	.29	.28	.57	
Finish, taped & finished joints	1.000	S.F.	.008	.04	.28	.32	
Corners, taped & finished, 32 L.F. per 12′ x 12′ room	.083	L.F.	.001	.01	.05	.06	
Painting, primer & 2 coats	1.000	S.F.	.011	.15	.32	.47	
Trim, baseboard, painted	.125	L.F.	.006	.19	.21	.40	
TOTAL			.034	.68	1.14	1.82	
FIRE RESISTANT, 1/2″ SHEETROCK, TAPED & FINISHED							
Drywall, 1/2″ thick, fire resistant	1.000	S.F.	.008	.26	.28	.54	
Finish, taped & finished joints	1.000	S.F.	.008	.04	.28	.32	
Corners, taped & finished, 32 L.F. per 12′ x 12′ room	.083	L.F.	.001	.01	.05	.06	
Painting, primer & 2 coats	1.000	S.F.	.011	.15	.32	.47	
Trim, baseboard, painted	.125	L.F.	.006	.19	.21	.40	
TOTAL			.034	.65	1.14	1.79	
WATER RESISTANT, 1/2″ SHEETROCK, TAPED & FINISHED							
Drywall, 1/2″ thick, water resistant	1.000	S.F.	.008	.37	.28	.65	
Finish, taped & finished joints	1.000	S.F.	.008	.04	.28	.32	
Corners, taped & finished, 32 L.F. per 12′ x 12′ room	.083	L.F.	.001	.01	.05	.06	
Painting, primer & 2 coats	1.000	S.F.	.011	.15	.32	.47	
Trim, baseboard, painted	.125	L.F.	.006	.19	.21	.40	
TOTAL			.034	.76	1.14	1.90	
THINCOAT, SKIM-COAT, ON 1/2″ BACKER DRYWALL							
Drywall, 1/2″ thick, thincoat backer	1.000	S.F.	.008	.29	.28	.57	
Thincoat plaster	1.000	S.F.	.011	.08	.35	.43	
Corners, taped & finished, 32 L.F. per 12′ x 12′ room	.083	L.F.	.001	.01	.05	.06	
Painting, primer & 2 coats	1.000	S.F.	.011	.15	.32	.47	
Trim, baseboard, painted	.125	L.F.	.006	.19	.21	.40	
TOTAL			.037	.72	1.21	1.93	
5/8″ SHEETROCK, TAPED & FINISHED							
Drywall, 5/8″ thick, standard	1.000	S.F.	.008	.32	.28	.60	
Finish, taped & finished joints	1.000	S.F.	.008	.04	.28	.32	
Corners, taped & finished, 32 L.F. per 12′ x 12′ room	.083	L.F.	.001	.01	.05	.06	
Painting, primer & 2 coats	1.000	S.F.	.011	.15	.32	.47	
Trim, baseboard, painted	.125	L.F.	.006	.19	.21	.40	
TOTAL			.034	.71	1.14	1.85	

Drywall & Thincoat Wall Systems

System Description	QUAN.	UNIT	LABOR-HOURS	COST PER S.F. MAT.	COST PER S.F. INST.	COST PER S.F. TOTAL	____S.F. EXTENSIONS
3/8" SHEETROCK, TAPED & FINISHED							
Drywall, 3/8" thick, standard	1.000	S.F.	.008	.29	.28	.57	
Finish, taped & finished joints	1.000	S.F.	.008	.04	.28	.32	
Corners, taped & finished, 32 L.F. per 12' x 12' room	.083	L.F.	.001	.01	.05	.06	
Painted, primer & 2 coats	1.000	S.F.	.011	.15	.32	.47	
Trim, baseboard, painted	.125	L.F.	.006	.19	.21	.40	
TOTAL			.034	.68	1.14	1.82	
THINCOAT, SKIM-COAT, ON 3/8" BACKER DRYWALL							
Drywall, 3/8" thick, thincoat backer	1.000	S.F.	.008	.29	.28	.57	
Thincoat, plaster	1.000	S.F.	.011	.08	.35	.43	
Corners, taped & finished, 32 L.F. per 12' x 12' room	.083	L.F.	.001	.01	.05	.06	
Painting, primer & 2 coats	1.000	S.F.	.011	.15	.32	.47	
Trim, baseboard, painted	.125	L.F.	.006	.19	.21	.40	
TOTAL			.037	.72	1.21	1.93	
WATER-RESISTANT, 3/8" SHEETROCK, TAPED & FINISHED							
Drywall, 3/8" thick, water-resistant	1.000	S.F.	.008	.29	.28	.57	
Finish, taped & finished joints	1.000	S.F.	.008	.04	.28	.32	
Corners, taped & finished, 32 L.F. per 12' x 12' room	.083	L.F.	.001	.01	.05	.06	
Painting, primer & 2 coats	1.000	S.F.	.011	.15	.32	.47	
Trim, baseboard, painted	.125	L.F.	.006	.19	.21	.40	
TOTAL			.034	.68	1.14	1.82	

The costs in this system are based on a square foot of wall.
Do not deduct for openings.

Drywall & Thincoat Wall System Components

Component Description	QUAN.	UNIT	LABOR-HOURS	COST PER S.F. MAT.	COST PER S.F. INST.	COST PER S.F. TOTAL
Drywall - sheetrock						
Drywall-sheetrock, 1/2" thick, standard	1.000	S.F.	.008	.29	.28	.57
Fire resistant	1.000	S.F.	.008	.26	.28	.54
Water resistant	1.000	S.F.	.008	.37	.28	.65
Paneling, not including furring or trim						
Plywood, prefinished, 1/4" thick, 4' x 8' sheets, vert. grooves						
Birch faced, minimum	1.000	S.F.	.032	.81	1.11	1.92
Average	1.000	S.F.	.038	1.23	1.33	2.56
Maximum	1.000	S.F.	.046	1.80	1.59	3.39
Mahogany, African	1.000	S.F.	.040	2.31	1.39	3.70
Philippine (lauan)	1.000	S.F.	.032	.99	1.11	2.10
Oak or cherry, minimum	1.000	S.F.	.032	1.94	1.11	3.05
Maximum	1.000	S.F.	.040	2.97	1.39	4.36
Rosewood	1.000	S.F.	.050	4.21	1.74	5.95
Teak	1.000	S.F.	.040	2.97	1.39	4.36
Chestnut	1.000	S.F.	.043	4.39	1.48	5.87
Pecan	1.000	S.F.	.040	1.89	1.39	3.28
Walnut, minimum	1.000	S.F.	.032	2.53	1.11	3.64
Maximum	1.000	S.F.	.040	4.80	1.39	6.19

Drywall & Thincoat Ceiling Systems

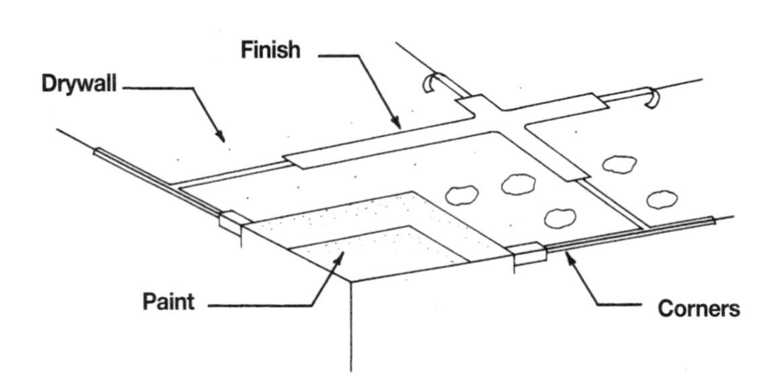

System Description	QUAN.	UNIT	LABOR-HOURS	COST PER S.F. MAT.	COST PER S.F. INST.	COST PER S.F. TOTAL	____ S.F. EXTENSIONS
1/2" SHEETROCK, TAPED & FINISHED							
Drywall, 1/2" thick, standard	1.000	S.F.	.008	.29	.28	.57	
Finish, taped & finished	1.000	S.F.	.008	.04	.28	.32	
Corners, taped & finished, 12' x 12' room	.333	L.F.	.005	.02	.17	.19	
Paint, primer & 2 coats	1.000	S.F.	.011	.15	.32	.47	
TOTAL			.032	.50	1.05	1.55	
THINCOAT, SKIM COAT ON 1/2" BACKER DRYWALL							
Drywall, 1/2" thick, thincoat backer	1.000	S.F.	.008	.29	.28	.57	
Thincoat plaster	1.000	S.F.	.011	.08	.35	.43	
Corners, taped & finished, 12' x 12' room	.333	L.F.	.005	.02	.17	.19	
Paint, primer & 2 coats	1.000	S.F.	.011	.15	.32	.47	
TOTAL			.035	.54	1.12	1.66	
FIRE RESISTANT, 1/2" SHEETROCK, TAPED & FINISHED							
Drywall, 1/2" thick, fire resistant	1.000	S.F.	.008	.26	.28	.54	
Finish, taped & finished	1.000	S.F.	.008	.04	.28	.32	
Corners, taped & finished, 12' x 12' room	.333	L.F.	.005	.02	.17	.19	
Paint, primer & 2 coats	1.000	S.F.	.011	.15	.32	.47	
TOTAL			.032	.47	1.05	1.52	
WATER-RESISTANT SHEETROCK, 1/2" THICK, TAPED & FINISHED							
Drywall, 1/2" thick, water-resistant	1.000	S.F.	.008	.37	.28	.65	
Finish, taped & finished	1.000	S.F.	.008	.04	.28	.32	
Corners, taped & finished, 12' x 12' room	.333	L.F.	.005	.02	.17	.19	
Paint, primer & 2 coats	1.000	S.F.	.011	.15	.32	.47	
TOTAL			.032	.58	1.05	1.63	
5/8" SHEETROCK, TAPED & FINISHED							
Drywall, 5/8" thick, standard	1.000	S.F.	.008	.32	.28	.60	
Finish, taped & finished	1.000	S.F.	.008	.04	.28	.32	
Corners, taped & finished, 12' x 12' room	.333	L.F.	.005	.02	.17	.19	
Paint, primer & 2 coats	1.000	S.F.	.011	.15	.32	.47	
TOTAL			.032	.53	1.05	1.58	

Drywall & Thincoat Ceiling Systems

System Description	QUAN.	UNIT	LABOR-HOURS	COST PER S.F. MAT.	INST.	TOTAL	S.F. EXTENSIONS
3/8" SHEETROCK, TAPED & FINISHED							
Drywall, 3/8" thick, standard	1.000	S.F.	.008	.29	.28	.57	
Finish, taped & finished	1.000	S.F.	.008	.04	.28	.32	
Corners, taped & finished, 12' x 12' room	.330	L.F.	.005	.02	.17	.19	
Paint, primer & 2 coats	1.000	S.F.	.011	.15	.32	.47	
TOTAL			.032	.50	1.05	1.55	
THINCOAT, SKIM-COAT ON 3/8" BACKER DRYWALL							
Drywall, 3/8" thick, thincoat backer	1.000	S.F.	.008	.29	.28	.57	
Thincoat plaster	1.000	S.F.	.011	.08	.35	.43	
Corners, taped & finished, 12' x 12' room	.330	L.F.	.005	.02	.17	.19	
Paint, primer & 2 coats	1.000	S.F.	.011	.15	.32	.47	
TOTAL			.035	.54	1.12	1.66	
WATER-RESISTANT, 3/8" SHEETROCK, TAPED & FINISHED							
Drywall, 3/8" thick, water-resistant	1.000	S.F.	.008	.29	.28	.57	
Finish, taped & finished	1.000	S.F.	.008	.04	.28	.32	
Corners, taped & finished, 12' x 12' room	.330	L.F.	.005	.02	.17	.19	
Paint, primer & 2 coats	1.000	S.F.	.011	.15	.32	.47	
TOTAL			.032	.50	1.05	1.55	

The costs in this system are based on a square foot of ceiling.

Drywall & Thincoat Ceiling System Components

Component Description	QUAN.	UNIT	LABOR-HOURS	COST PER S.F. MAT.	INST.	TOTAL
Drywall - sheetrock						
5/8" thick, standard	1.000	S.F.	.008	.32	.28	.60
Fire resistant	1.000	S.F.	.008	.32	.28	.60
Water resistant	1.000	S.F.	.008	.47	.28	.75
Drywall backer for thincoat system, 1/2" thick	1.000	S.F.	.016	.55	.56	1.11
5/8" thick	1.000	S.F.	.016	.58	.56	1.14
Corners taped & finished						
Corners taped & finished, 4' x 4' room	1.000	L.F.	.015	.07	.51	.58
6' x 6' room	.667	L.F.	.010	.05	.34	.39
10' x 10' room	.400	L.F.	.006	.03	.20	.23
12' x 12' room	.333	L.F.	.005	.02	.17	.19
16' x 16' room	.250	L.F.	.003	.01	.10	.11
Thincoat system, 4' x 4' room	1.000	L.F.	.011	.08	.35	.43
6' x 6' room	.667	L.F.	.007	.05	.23	.28
10' x 10' room	.400	L.F.	.004	.03	.14	.17
12' x 12' room	.333	L.F.	.004	.03	.12	.15
16' x 16' room	.250	L.F.	.002	.01	.06	.07
Tile						
Tile, ceramic adhesive thin set, 4 1/4" x 4 1/4" tiles	1.000	S.F.	.084	2.30	2.43	4.73
6" x 6" tiles	1.000	S.F.	.080	2.78	2.30	5.08
Pregrouted sheets	1.000	S.F.	.067	4.57	1.92	6.49

Plaster & Stucco Wall Systems

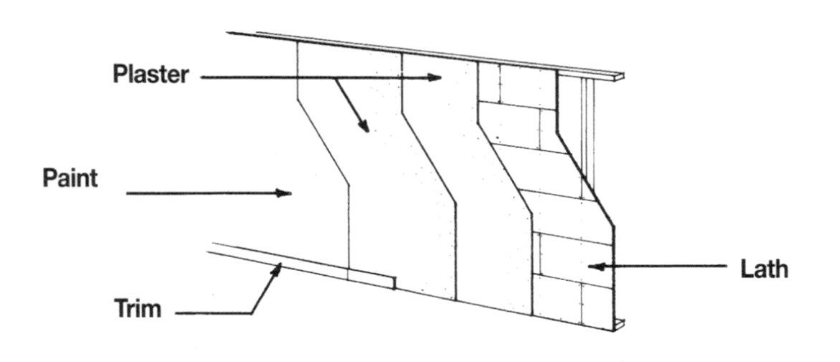

System Description	QUAN.	UNIT	LABOR-HOURS	COST PER S.F. MAT.	COST PER S.F. INST.	COST PER S.F. TOTAL	____S.F. EXTENSIONS
PLASTER ON GYPSUM LATH							
Plaster, gypsum or perlite, 2 coats	1.000	S.F.	.053	.44	1.69	2.13	
Lath, 3/8'' gypsum	1.000	S.F.	.010	.43	.34	.77	
Corners, expanded metal, 32 L.F. per 12' x 12' room	.083	L.F.	.002	.01	.06	.07	
Painting, primer & 2 coats	1.000	S.F.	.011	.15	.32	.47	
Trim, baseboard, painted	.125	L.F.	.006	.19	.21	.40	
TOTAL			.082	1.22	2.62	3.84	
PLASTER ON GYPSUM LATH, 3 COATS							
Plaster, gypsum or perlite, 3 coats	1.000	S.F.	.065	.62	2.04	2.66	
Lath, 3/8'' gypsum	1.000	S.F.	.010	.43	.34	.77	
Corners, expanded metal, 32 L.F. per 12' x 12' room	.083	L.F.	.002	.01	.06	.07	
Painting, primer & 2 coats	1.000	S.F.	.011	.15	.32	.47	
Trim, baseboard, painted	.125	L.F.	.006	.19	.21	.40	
TOTAL			.094	1.40	2.97	4.37	
PLASTER ON METAL LATH							
Plaster, gypsum or perlite, 2 coats	1.000	S.F.	.053	.44	1.69	2.13	
Lath, 2.5 Lb. diamond, metal	1.000	S.F.	.010	.17	.34	.51	
Corners, expanded metal, 32 L.F. per 12' x 12' room	.083	L.F.	.002	.01	.06	.07	
Painting, primer & 2 coats	1.000	S.F.	.011	.15	.32	.47	
Trim, baseboard, painted	.125	L.F.	.006	.19	.21	.40	
TOTAL			.082	.96	2.62	3.58	
PLASTER ON METAL LATH, 3 COATS							
Plaster, gypsum or perlite, 3 coats	1.000	S.F.	.065	.62	2.04	2.66	
Lath, 2.5 Lb. diamond, metal	1.000	S.F.	.010	.17	.34	.51	
Corners, expanded metal, 32 L.F. per 12' x 12' room	.083	L.F.	.002	.01	.06	.07	
Painting, primer & 2 coats	1.000	S.F.	.011	.15	.32	.47	
Trim, baseboard, painted	.125	L.F.	.006	.19	.21	.40	
TOTAL			.094	1.14	2.97	4.11	

Plaster & Stucco Wall Systems

System Description	QUAN.	UNIT	LABOR-HOURS	COST PER S.F. MAT.	INST.	TOTAL	____ S.F. EXTENSIONS
STUCCO ON METAL LATH							
Stucco, 2 coats	1.000	S.F.	.041	.23	1.30	1.53	
Lath, 2.5 Lb. diamond, metal	1.000	S.F.	.010	.17	.34	.51	
Corners, expanded metal, 32 L.F. per 12′ x 12′ room	.083	L.F.	.002	.01	.06	.07	
Painting, primer & 2 coats	1.000	S.F.	.011	.15	.32	.47	
Trim, baseboard, painted	.125	L.F.	.006	.19	.21	.40	
TOTAL			.070	.75	2.23	2.98	
STUCCO ON METAL LATH, 3 COATS							
Stucco, 3 coats	1.000	S.F.	.102	.40	3.25	3.65	
Lath, 2.5 Lb. diamond, metal	1.000	S.F.	.010	.17	.34	.51	
Corners, expanded metal, 32 L.F. per 12′ x 12′ room	.083	L.F.	.002	.01	.06	.07	
Painting, primer & 2 coats	1.000	S.F.	.011	.15	.32	.47	
Trim, baseboard, painted	.125	L.F.	.006	.19	.21	.40	
TOTAL			.131	.92	4.18	5.10	

The costs in these systems are based on a per square foot of wall area.
Do not deduct for openings.

Plaster & Stucco Wall System Components

Component Description	QUAN.	UNIT	LABOR-HOURS	COST PER S.F. MAT.	INST.	TOTAL
Lath						
Lath, gypsum, standard, 3/8″ thick	1.000	S.F.	.010	.43	.34	.77
1/2″ thick	1.000	S.F.	.013	.44	.41	.85
Fire resistant, 3/8″ thick	1.000	S.F.	.013	.43	.41	.84
1/2″ thick	1.000	S.F.	.014	.48	.44	.92
Metal, diamond, 2.5 Lb.	1.000	S.F.	.010	.17	.34	.51
3.4 Lb.	1.000	S.F.	.012	.25	.38	.63
Rib, 2.75 Lb.	1.000	S.F.	.012	.25	.38	.63
3.4 Lb.	1.000	S.F.	.013	.36	.41	.77
Corners						
Corners, expanded metal, 32 L.F. per 4′ x 4′ room	.250	L.F.	.005	.03	.18	.21
6′ x 6′ room	.110	L.F.	.002	.01	.08	.09
10′ x 10′ room	.100	L.F.	.002	.01	.07	.08
16′ x 16′ room	.063	L.F.	.001	.01	.04	.05
Painting						
Painting, primer & 1 coats	1.000	S.F.	.008	.10	.25	.35
Wallpaper, low price double roll	1.000	S.F.	.013	.31	.39	.70
Medium price double roll	1.000	S.F.	.015	.67	.46	1.13
High price double roll	1.000	S.F.	.018	1.61	.57	2.18
Tile, ceramic thin set, 4-1/4″ x 4-1/4″ tiles	1.000	S.F.	.084	2.30	2.43	4.73
6″ x 6″ tiles	1.000	S.F.	.080	2.78	2.30	5.08
Pregrouted sheets	1.000	S.F.	.067	4.57	1.92	6.49

Plaster & Stucco Ceiling Systems

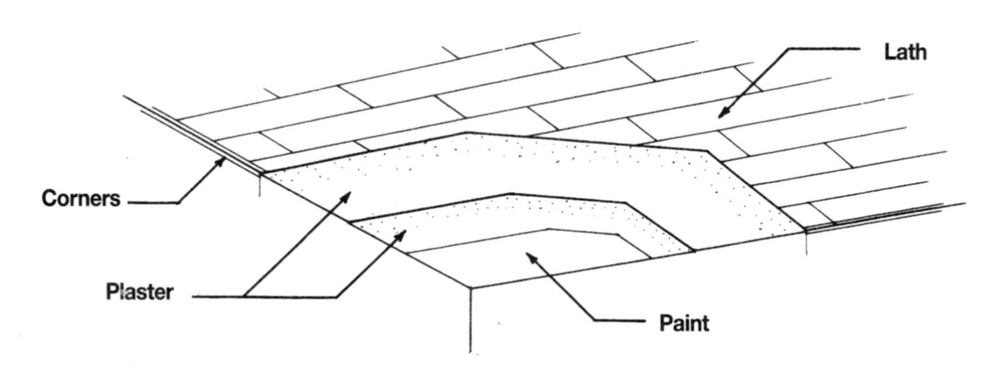

System Description	QUAN.	UNIT	LABOR-HOURS	COST PER S.F. MAT.	COST PER S.F. INST.	COST PER S.F. TOTAL	____S.F. EXTENSIONS
PLASTER ON GYPSUM LATH							
Plaster, gypsum or perlite, 2 coats	1.000	S.F.	.061	.42	1.92	2.34	
Lath, 3/8″ gypsum	1.000	S.F.	.014	.43	.47	.90	
Corners, expanded metal, 12′ x 12′ room	.330	L.F.	.007	.04	.23	.27	
Painting, primer & 2 coats	1.000	S.F.	.011	.15	.32	.47	
TOTAL			.093	1.04	2.94	3.98	
PLASTER ON GYPSUM LATH, 3 COATS							
Plaster, gypsum or perlite, 3 coats	1.000	S.F.	.065	.62	2.04	2.66	
Lath, 3/8″ gypsum	1.000	S.F.	.014	.43	.47	.90	
Corners, expanded metal, 12′ x 12′ room	.330	L.F.	.007	.04	.23	.27	
Painting, primer & 2 coats	1.000	S.F.	.011	.15	.32	.47	
TOTAL			.097	1.24	3.06	4.30	
PLASTER ON METAL LATH							
Plaster, gypsum or perlite, 2 coats	1.000	S.F.	.061	.42	1.92	2.34	
Lath, 2.5 Lb. diamond, metal	1.000	S.F.	.012	.17	.38	.55	
Corners, expanded metal, 12′ x 12′ room	.330	L.F.	.007	.04	.23	.27	
Painting, primer & 2 coats	1.000	S.F.	.011	.15	.32	.47	
TOTAL			.091	.78	2.85	3.63	
PLASTER ON METAL LATH, 3 COATS							
Plaster, gypsum or perlite, 3 coats	1.000	S.F.	.065	.62	2.04	2.66	
Lath, 2.5 Lb. diamond, metal	1.000	S.F.	.012	.17	.38	.55	
Corners, expanded metal, 12′ x 12′ room	.330	L.F.	.007	.04	.23	.27	
Painting, primer & 2 coats	1.000	S.F.	.011	.15	.32	.47	
TOTAL			.095	.98	2.97	3.95	
STUCCO ON GYPSUM LATH							
Stucco, 2 coats	1.000	S.F.	.041	.23	1.30	1.53	
Lath, 3/8″ gypsum	1.000	S.F.	.014	.43	.47	.90	
Corners, expanded metal, 12′ x 12′ room	.330	L.F.	.007	.04	.23	.27	
Painting, primer & 2 coats	1.000	S.F.	.011	.15	.32	.47	
TOTAL			.073	.85	2.32	3.17	

Plaster & Stucco Ceiling Systems

System Description	QUAN.	UNIT	LABOR-HOURS	COST PER S.F.			____S.F. EXTENSIONS
				MAT.	INST.	TOTAL	
STUCCO ON GYPSUM LATH, 3 COATS							
Stucco, 3 coats	1.000	S.F.	.102	.40	3.25	3.65	
Lath, 3/8" gypsum	1.000	S.F.	.014	.43	.47	.90	
Corners, expanded metal, 12' x 12' room	.330	L.F.	.007	.04	.23	.27	
Painting, primer & 2 coats	1.000	S.F.	.011	.15	.32	.47	
TOTAL			.134	1.02	4.27	5.29	
STUCCO ON METAL LATH							
Stucco, 2 coats	1.000	S.F.	.041	.23	1.30	1.53	
Lath, 2.5 Lb. diamond, metal	1.000	S.F.	.012	.17	.38	.55	
Corners, expanded metal, 12' x 12' room	.330	L.F.	.007	.04	.23	.27	
Painting, primer & 2 coats	1.000	S.F.	.011	.15	.32	.47	
TOTAL			.071	.59	2.23	2.82	
STUCCO ON METAL LATH, 3 COATS							
Stucco, 3 coats	1.000	S.F.	.102	.40	3.25	3.65	
Lath, 2.5 Lb. diamond, metal	1.000	S.F.	.012	.17	.38	.55	
Corners, expanded metal, 12' x 12' room	.330	L.F.	.007	.04	.23	.27	
Painting, primer & 2 coats	1.000	S.F.	.011	.15	.32	.47	
TOTAL			.132	.76	4.18	4.94	

The costs in these systems are based on a square foot of ceiling area.

Plaster & Stucco Ceiling System Components

Component Description	QUAN.	UNIT	LABOR-HOURS	COST PER S.F.		
				MAT.	INST.	TOTAL
Lath						
Lath, gypsum, standard, 3/8" thick	1.000	S.F.	.014	.43	.47	.90
1/2" thick	1.000	S.F.	.015	.43	.49	.92
Fire resistant, 3/8" thick	1.000	S.F.	.017	.43	.54	.97
1/2" thick	1.000	S.F.	.018	.48	.57	1.05
Metal, diamond, 2.5 Lb.	1.000	S.F.	.012	.17	.38	.55
3.4 Lb.	1.000	S.F.	.015	.25	.48	.73
Rib, 2.75 Lb.	1.000	S.F.	.012	.25	.38	.63
3.4 Lb.	1.000	S.F.	.013	.36	.41	.77
Corners						
Corners expanded metal, 4' x 4' room	1.000	L.F.	.020	.11	.70	.81
6' x 6' room	.667	L.F.	.013	.07	.47	.54
10' x 10' room	.400	L.F.	.008	.04	.28	.32
16' x 16' room	.250	L.F.	.004	.02	.13	.15
Painting						
Painting, primer & 1 coat	1.000	S.F.	.008	.10	.25	.35

Interior Door Systems

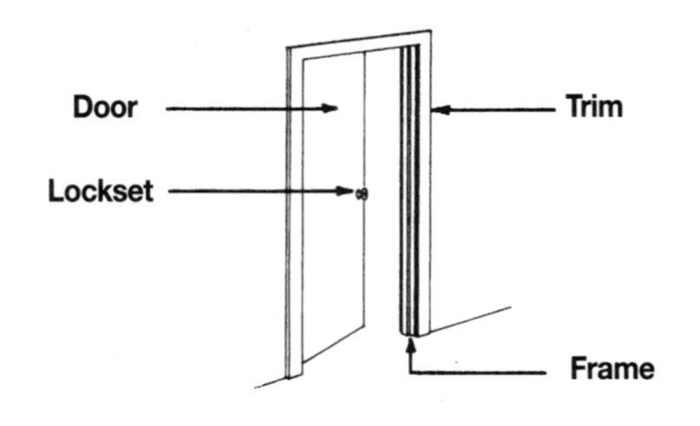

System Description	QUAN.	UNIT	LABOR-HOURS	COST EACH MAT.	COST EACH INST.	COST EACH TOTAL	____EA. EXTENSIONS
LAUAN, DOOR, HOLLOW CORE, 1-3/8", 2'-0" X 6'-8"							
Door, flush lauan, hollow core, 2'-0" x 6'-8" high	1.000	Ea.	.889	27.50	31.00	58.50	
Frame, pine, 4-5/8" jamb	16.000	L.F.	.683	91.20	23.68	114.88	
Trim, casing, painted	32.000	L.F.	1.461	28.48	49.28	77.76	
Butt hinges, chrome, 3-1/2" x 3-1/2"	1.500	Pr.		24.08		24.08	
Lockset, passage	1.000	Ea.	.500	13.85	17.40	31.25	
Paint, door & frame, primer & 2 coats	2.000	Face	5.547	13.78	171.20	184.98	
TOTAL			9.080	198.89	292.56	491.45	
BIRCH, DOOR, HOLLOW CORE, 1-3/8", 2'-0" X 6'-8"							
Door, flush, birch, hollow core, 2'-0" x 6'-8" high	1.000	Ea.	.889	39.50	31.00	70.50	
Frame, pine, 4-5/8" jamb	16.000	L.F.	.683	91.20	23.68	114.88	
Trim, casing, painted	32.000	L.F.	1.461	28.48	49.28	77.76	
Butt hinges, chrome, 3-1/2" x 3-1/2"	1.500	Pr.		24.08		24.08	
Lockset, passage	1.000	Ea.	.500	13.85	17.40	31.25	
Paint, door & frame, primer & 2 coats	2.000	Face	5.547	13.78	171.20	184.98	
TOTAL			9.080	210.89	292.56	503.45	
PINE, PANELED, 1-3/8" THICK, 2'-0" X 6'-8"							
Door, pine, raised panel, 2'-0" wide x 6'-8" high	1.000	Ea.	.889	128.00	31.00	159.00	
Frame, pine, 4-5/8" jamb	16.000	L.F.	.683	91.20	23.68	114.88	
Trim, casing, painted	32.000	L.F.	1.461	28.48	49.28	77.76	
Butt hinges, bronze, 3-1/2" x 3-1/2"	1.500	Pr.		24.08		24.08	
Lockset, passage	1.000	Ea.	.500	13.85	17.40	31.25	
Paint, door & frame, primer & 2 coats	2.000	Face	5.547	13.78	171.20	184.98	
TOTAL			9.080	299.39	292.56	591.95	
LAUAN, FLUSH DOOR, HOLLOW CORE							
Door, flush, lauan, hollow core, 2'-8" wide x 6'-8" high	1.000	Ea.	.889	32.50	31.00	63.50	
Frame, pine, 4-5/8" jamb	17.000	L.F.	.725	96.90	25.16	122.06	
Trim, casing, painted	34.000	L.F.	1.473	32.64	49.98	82.62	
Butt hinges, chrome, 3-1/2" x 3-1/2"	1.500	Pr.		24.08		24.08	
Lockset, passage	1.000	Ea.	.500	13.85	17.40	31.25	
Paint, door & frame, primer & 2 coats	2.000	Face	3.886	12.08	120.00	132.08	
TOTAL			7.473	212.05	243.54	455.59	

Interior Door Systems

System Description	QUAN.	UNIT	LABOR-HOURS	COST EACH MAT.	COST EACH INST.	COST EACH TOTAL	____EA. EXTENSIONS
BIRCH, FLUSH DOOR, HOLLOW CORE							
Door, flush, birch, hollow core, 2'-8" wide x 6'-8" high	1.000	Ea.	.889	46.00	31.00	77.00	
Frame, pine, 4-5/8" jamb	17.000	L.F.	.725	96.90	25.16	122.06	
Trim, casing, painted	34.000	L.F.	1.473	32.64	49.98	82.62	
Butt hinges, chrome, 3-1/2" x 3-1/2"	1.500	Pr.		24.08		24.08	
Lockset, passage	1.000	Ea.	.500	13.85	17.40	31.25	
Paint, door & frame, primer & 2 coats	2.000	Face	3.886	12.08	120.00	132.08	
TOTAL			7.473	225.55	243.54	469.09	
RAISED PANEL, SOLID, PINE DOOR							
Door, pine, raised panel, 2'-8" wide x 6'-8" high	1.000	Ea.	.889	155.00	31.00	186.00	
Frame, pine, 4-5/8" jamb	17.000	L.F.	.725	96.90	25.16	122.06	
Trim, casing, painted	34.000	L.F.	1.473	32.64	49.98	82.62	
Butt hinges, bronze, 3-1/2" x 3-1/2"	1.500	Pr.		27.38		27.38	
Lockset, passage	1.000	Ea.	.500	13.85	17.40	31.25	
Paint, door & frame, primer & 2 coats	2.000	Face	8.000	16.72	247.00	263.72	
TOTAL			11.587	342.49	370.54	713.03	

The costs in these systems are based on a cost per each door.

Interior Door System Components

Component Description	QUAN.	UNIT	LABOR-HOURS	COST EACH MAT.	COST EACH INST.	COST EACH TOTAL
Doors						
Door, hollow core, lauan 1-3/8" thick, 6'-8" high x 1'-6" wide	1.000	Ea.	.889	25.50	31.00	56.50
2'-6" wide	1.000	Ea.	.889	31.00	31.00	62.00
3'-0" wide	1.000	Ea.	.941	34.50	33.00	67.50
Birch 1-3/8" thick, 6'-8" high x 1'-6" wide	1.000	Ea.	.889	33.00	31.00	64.00
2'-6" wide	1.000	Ea.	.889	46.00	31.00	77.00
3'-0" wide	1.000	Ea.	.941	52.00	33.00	85.00
Louvered pine 1-3/8" thick, 6'-8" high x 1'-6" wide	1.000	Ea.	.842	99.50	29.50	129.00
2'-0" wide	1.000	Ea.	.889	125.00	31.00	156.00
2'-6" wide	1.000	Ea.	.889	136.00	31.00	167.00
2'-8" wide	1.000	Ea.	.889	144.00	31.00	175.00
3'-0" wide	1.000	Ea.	.941	154.00	33.00	187.00
Paneled pine 1-3/8" thick, 6'-8" high x 1'-6" wide	1.000	Ea.	.842	111.00	29.50	140.50
2'-6" wide	1.000	Ea.	.889	143.00	31.00	174.00
3'-0" wide	1.000	Ea.	.941	165.00	33.00	198.00

Closet Door Systems

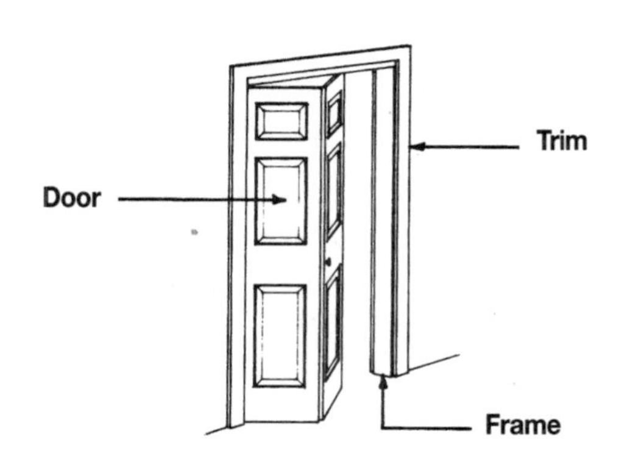

System Description	QUAN.	UNIT	LABOR-HOURS	COST EACH			____EA. EXTENSIONS
				MAT.	INST.	TOTAL	
BI-PASSING, FLUSH, LAUAN, HOLLOW CORE, 4′-0″ X 6′-8″							
Door, flush, lauan, hollow core, 4′-0″ x 6′-8″ opening	1.000	Ea.	1.333	171.00	46.50	217.50	
Frame, pine, 4-5/8″ jamb	18.000	L.F.	.768	102.60	26.64	129.24	
Trim, both sides, casing, painted	36.000	L.F.	2.086	33.12	69.12	102.24	
Paint, door & frame, primer & 2 coats	2.000	Face	3.886	12.08	120.00	132.08	
TOTAL			8.073	318.80	262.26	581.06	
BI-PASSING, PINE, PANELED, 4′-0″ X 6′-8″							
Door, pine, paneled, 4′-0″ x 6′-8″ opening	1.000	Ea.	1.333	370.00	46.50	416.50	
Frame, pine, 4-5/8″ jamb	18.000	L.F.	.768	102.60	26.64	129.24	
Trim, both sides, painted	36.000	L.F.	2.086	33.12	69.12	102.24	
Paint, door & frame, primer & 2 coats	2.000	Face	8.000	16.72	247.00	263.72	
TOTAL			12.187	522.44	389.26	911.70	
BI-PASSING, FLUSH, BIRCH, HOLLOW CORE, 6′-0″ X 6′-8″							
Door, flush, birch, hollow core, 6′-0″ x 6′-8″ opening	1.000	Ea.	1.600	233.00	55.50	288.50	
Frame, pine, 4-5/8″ jamb	19.000	L.F.	.811	108.30	28.12	136.42	
Trim, both sides, casing, painted	38.000	L.F.	2.203	34.96	72.96	107.92	
Paint, door & frame, primer & 2 coats	2.000	Face	4.857	15.10	150.00	165.10	
TOTAL			9.471	391.36	306.58	697.94	
BI-PASSING, PINE, PANELED, 6′-0″ X 6′-8″							
Door, pine, paneled, 6′-0″ x 6′-8″ opening	1.000	Ea.	1.600	460.00	55.50	515.50	
Frame, pine, 4-5/8″ jamb	19.000	L.F.	.811	108.30	28.12	136.42	
Trim, both sides, painted	38.000	L.F.	2.203	34.96	72.96	107.92	
Paint, door & frame, primer & 2 coats	2.000	Face	10.000	20.91	308.75	329.66	
TOTAL			14.614	624.17	465.33	1089.50	
BI-FOLD, PINE, PANELED, 3′-0″ X 6′-8″							
Door, pine, paneled, 3′-0″ x 6′-8″ opening	1.000	Ea.	1.231	124.00	43.00	167.00	
Frame, pine, 4-5/8″ jamb	17.000	L.F.	.725	96.90	25.16	122.06	
Trim, both sides, casing, painted	34.000	L.F.	1.971	31.28	65.28	96.56	
Paint, door & frame, primer & 2 coats	2.000	Face	8.000	16.72	247.00	263.72	
TOTAL			11.927	268.90	380.44	649.34	

Closet Door Systems

System Description	QUAN.	UNIT	LABOR-HOURS	COST EACH MAT.	COST EACH INST.	COST EACH TOTAL	____EA. EXTENSIONS
BI-FOLD, PINE, LOUVERED, 3'-0" X 6'-8"							
Door, pine, louvered, 3'-0" x 6'-8" opening	1.000	Ea.	1.231	124.00	43.00	167.00	
Frame, pine, 4-5/8" jamb	17.000	L.F.	.725	96.90	25.16	122.06	
Trim, both sides, painted	34.000	L.F.	1.971	31.28	65.28	96.56	
Paint, door & frame, primer & 2 coats	2.000	Face	6.000	12.55	185.25	197.80	
TOTAL			9.927	264.73	318.69	583.42	
BI-FOLD, PINE, LOUVERED, 6'-0" X 6'-8"							
Door, pine, louvered, 6'-0" x 6'-8" opening	1.000	Ea.	1.600	244.00	55.50	299.50	
Frame, pine, 4-5/8" jamb	19.000	L.F.	.811	108.30	28.12	136.42	
Trim, both sides, casing, painted	38.000	L.F.	2.203	34.96	72.96	107.92	
Paint, door & frame, primer & 2 coats	2.000	Face	10.000	20.91	308.75	329.66	
TOTAL			14.614	408.17	465.33	873.50	
BI-FOLD, FLUSH, BIRCH, 6'-0" X 6'-8"							
Door, flush, birch, hollow core, 6'-0" x 6'-8" opng.	1.000	Ea.	1.600	106.00	55.50	161.50	
Frame, pine, 4-5/8" jamb	19.000	L.F.	.811	108.30	28.12	136.42	
Trim, both sides, painted	38.000	L.F.	2.203	34.96	72.96	107.92	
Paint, door & frame, primer & 2 coats	2.000	Face	4.857	15.10	150.00	165.10	
TOTAL			9.471	264.36	306.58	570.94	

The costs in this system are based on a cost per each door.

Closet Door System Components

Component Description	QUAN.	UNIT	LABOR-HOURS	COST EACH MAT.	COST EACH INST.	COST EACH TOTAL
Doors						
Doors, bi-passing, pine, louvered, 4'-0" x 6'-8" opening	1.000	Ea.	1.333	390.00	46.50	436.50
6'-0" x 6'-8" opening	1.000	Ea.	1.600	475.00	55.50	530.50
Flush, birch, hollow core, 4'-0" x 6'-8" opening	1.000	Ea.	1.333	194.00	46.50	240.50
6'-0" x 6'-8" opening	1.000	Ea.	1.600	233.00	55.50	288.50
Flush, lauan, hollow core, 4'-0" x 6'-8" opening	1.000	Ea.	1.333	171.00	46.50	217.50
6'-0" x 6'-8" opening	1.000	Ea.	1.600	199.00	55.50	254.50
Bi-fold, pine, louvered, 3'-0" x 6'-8" opening	1.000	Ea.	1.231	124.00	43.00	167.00
6'-0" x 6'-8" opening	1.000	Ea.	1.600	244.00	55.50	299.50
Flush, birch, hollow core, 3'-0" x 6'-8" opening	1.000	Ea.	1.231	49.00	43.00	92.00
6'-0" x 6'-8" opening	1.000	Ea.	1.600	106.00	55.50	161.50
Flush, lauan, hollow core, 3'-0" x 6'8" opening	1.000	Ea.	1.231	149.00	43.00	192.00
6'-0" x 6'-8" opening	1.000	Ea.	1.600	310.00	55.50	365.50
Frames						
Frame pine, 3'-0" door, 3-5/8" deep	17.000	L.F.	.725	72.00	25.00	97.00
5-5/8" deep	17.000	L.F.	.725	90.00	25.00	115.00
4'-0" door, 3-5/8" deep	18.000	L.F.	.768	76.50	26.50	103.00
5-5/8" deep	18.000	L.F.	.768	95.50	26.50	122.00
6'-0" door, 3-5/8" deep	19.000	L.F.	.811	80.50	28.00	108.50
5-5/8" deep	19.000	L.F.	.811	101.00	28.00	129.00

Section Two

Construction Information

Provided for your use are illustrations together with the common terms used for the various components of framing and rough carpentry. The reason for this is that terms used are not universal. What someone in one part of the country might call a bottom plate, someone else might call a sole plate.

In this section are the common names used for various exterior wood framing systems, siding systems and roof framing systems. Illustrations and terminology are also shown for various interior systems such as partition framing systems, ceiling systems, and door and window systems.

Wood Framing Nomenclature

These graphics illustrate the common terms used for the various components of wood framing. Whether for residential (shown here) or commercial, the terminology is the same.

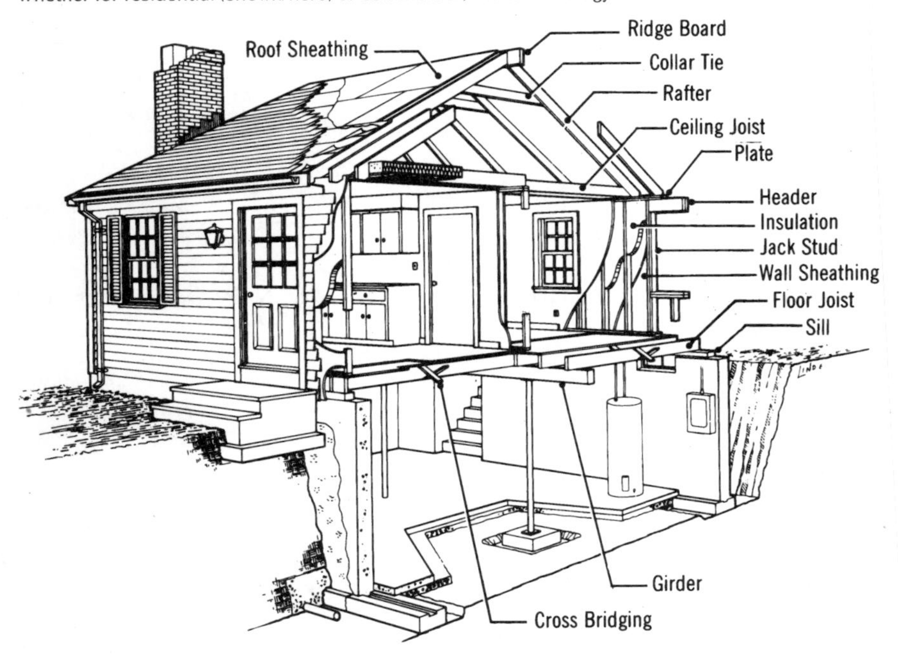

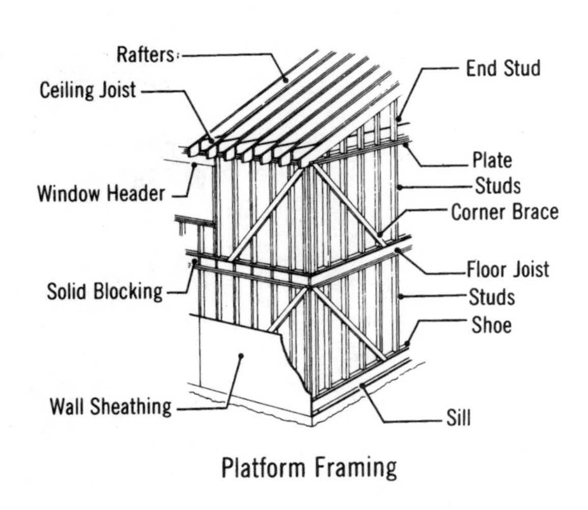

Platform Framing

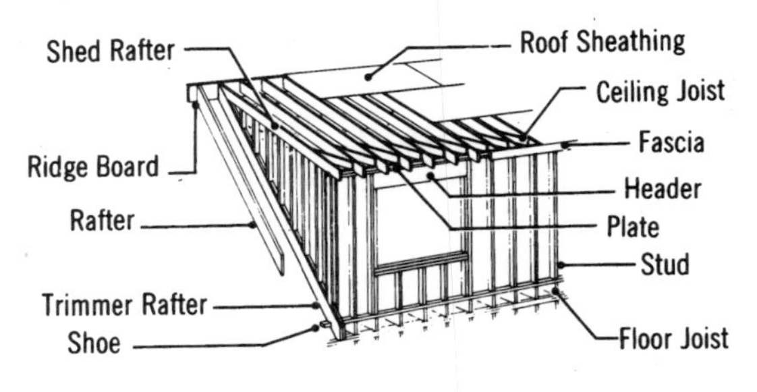

Shed Dormer Framing

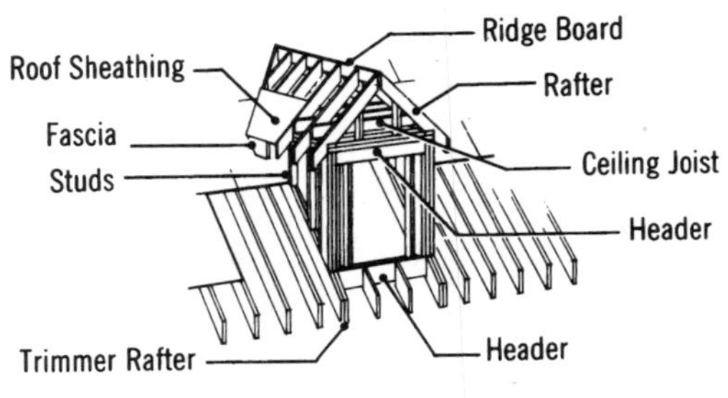

Gable Dormer Framing

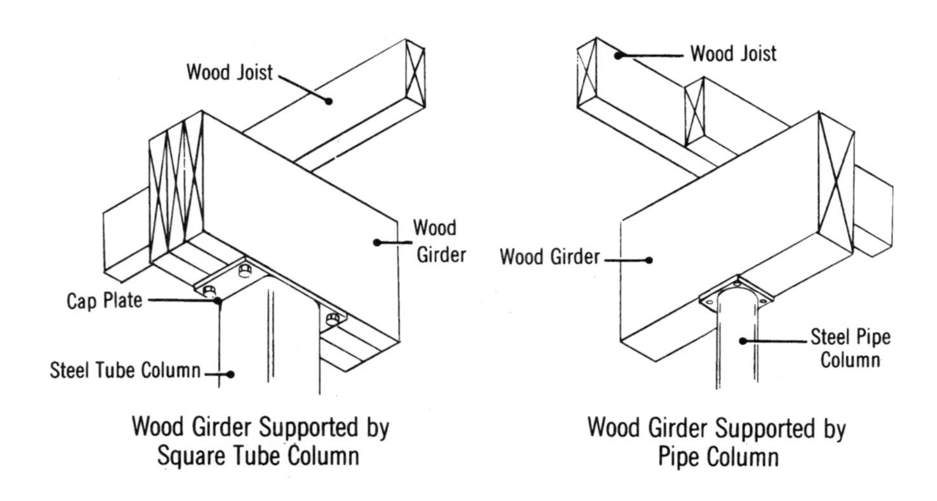

Wood Girder Supported by Square Tube Column

Wood Girder Supported by Pipe Column

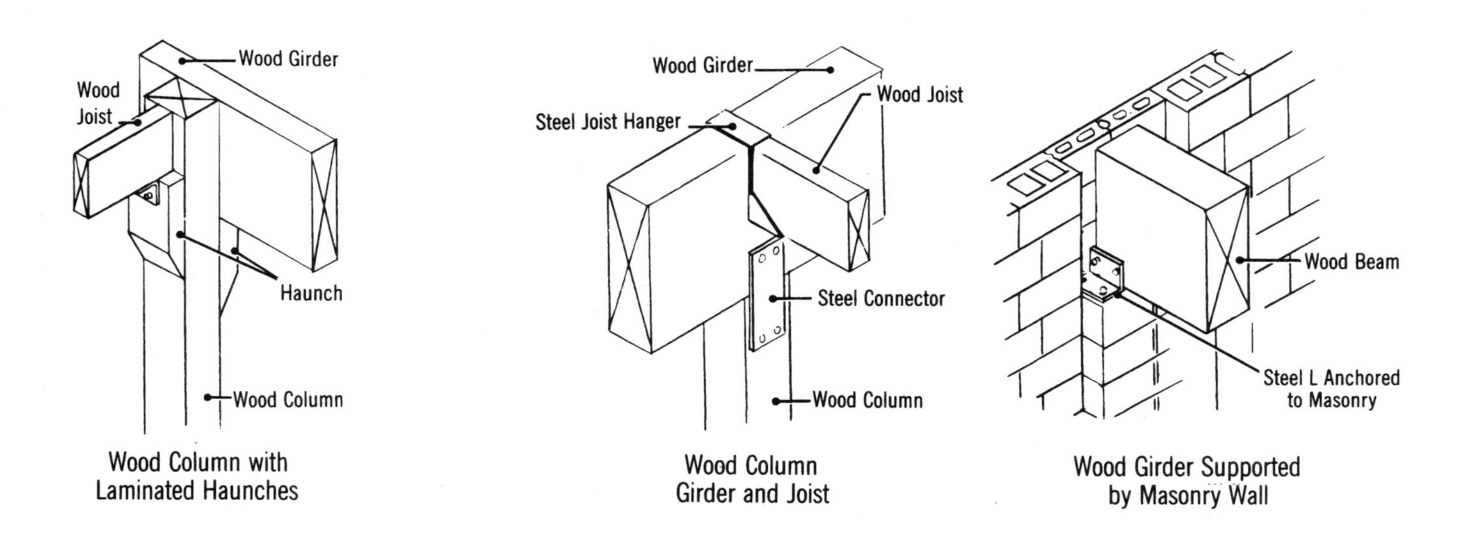

Wood Column with Laminated Haunches

Wood Column Girder and Joist

Wood Girder Supported by Masonry Wall

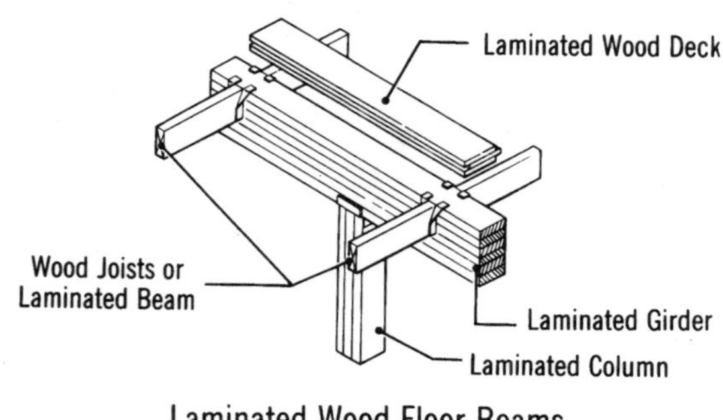

Laminated Wood Floor Beams

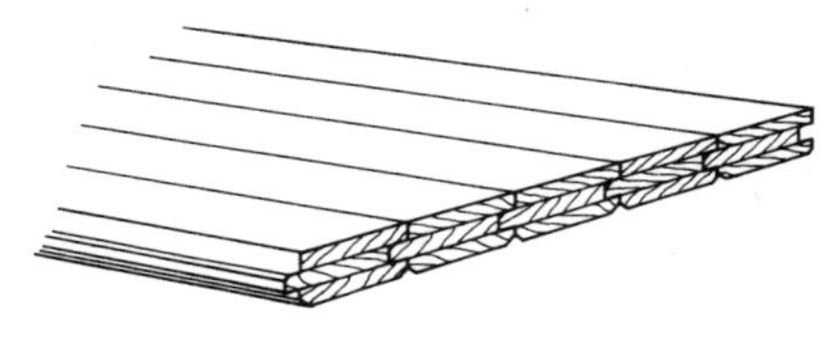

Laminated Wood Deck

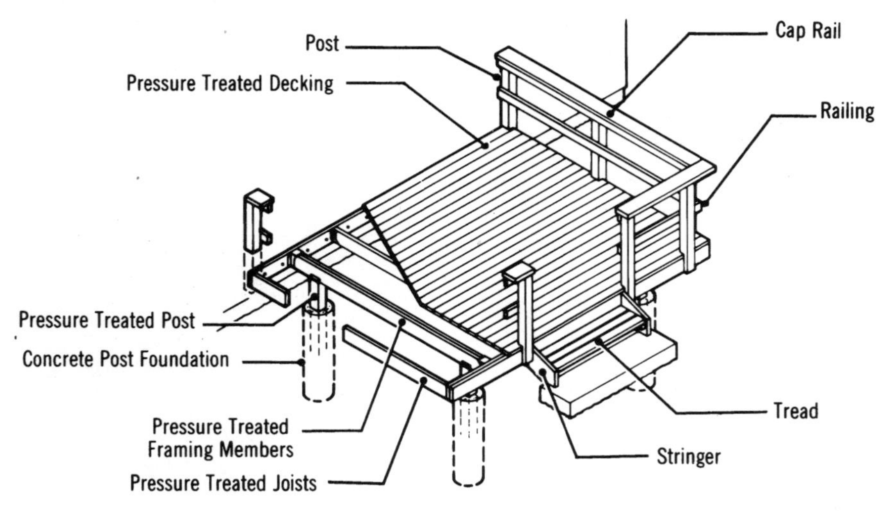

Wood Deck Construction

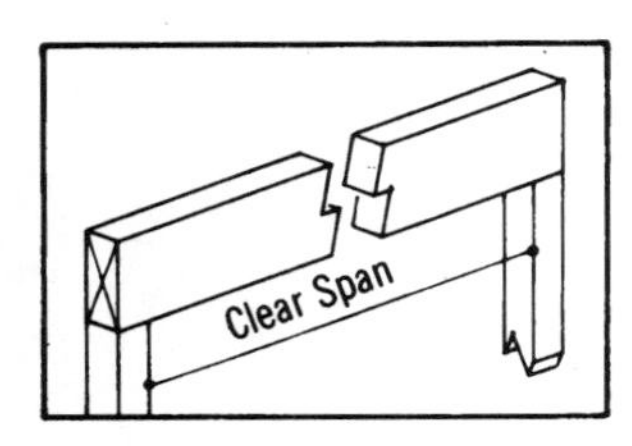

Single Beam

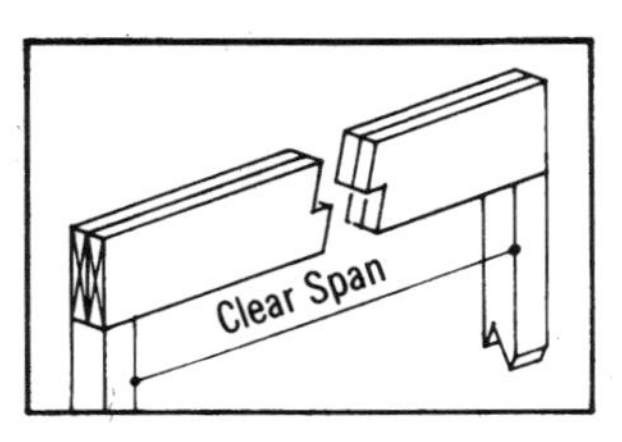

Double Beam

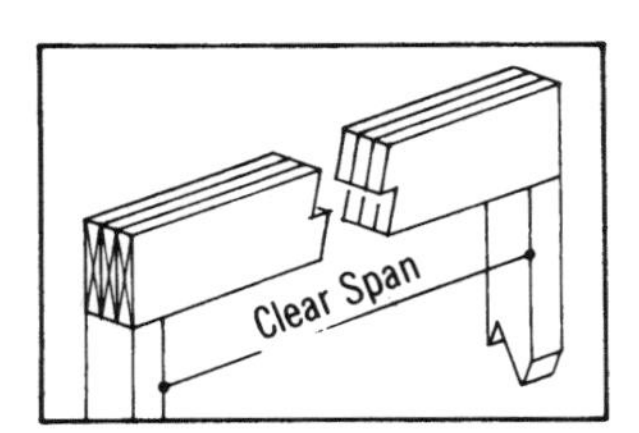

Triple Beam

Wood Beams and Columns

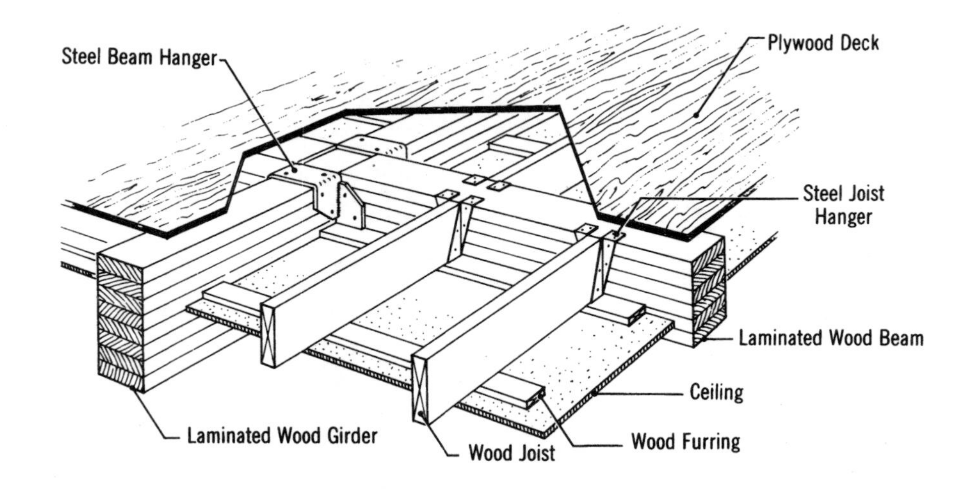

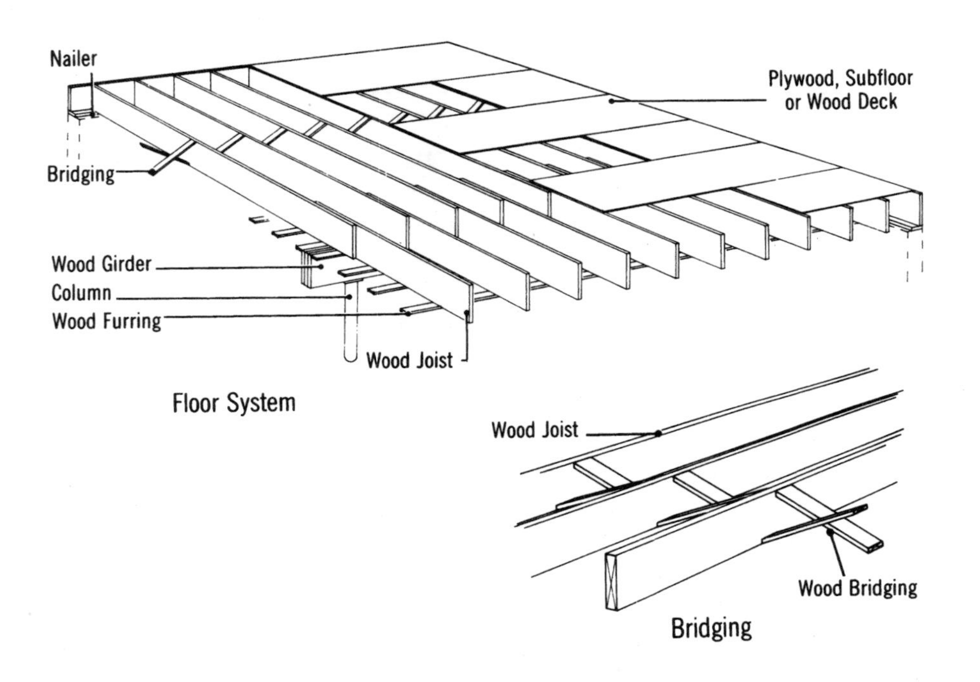

Floor System

Bridging

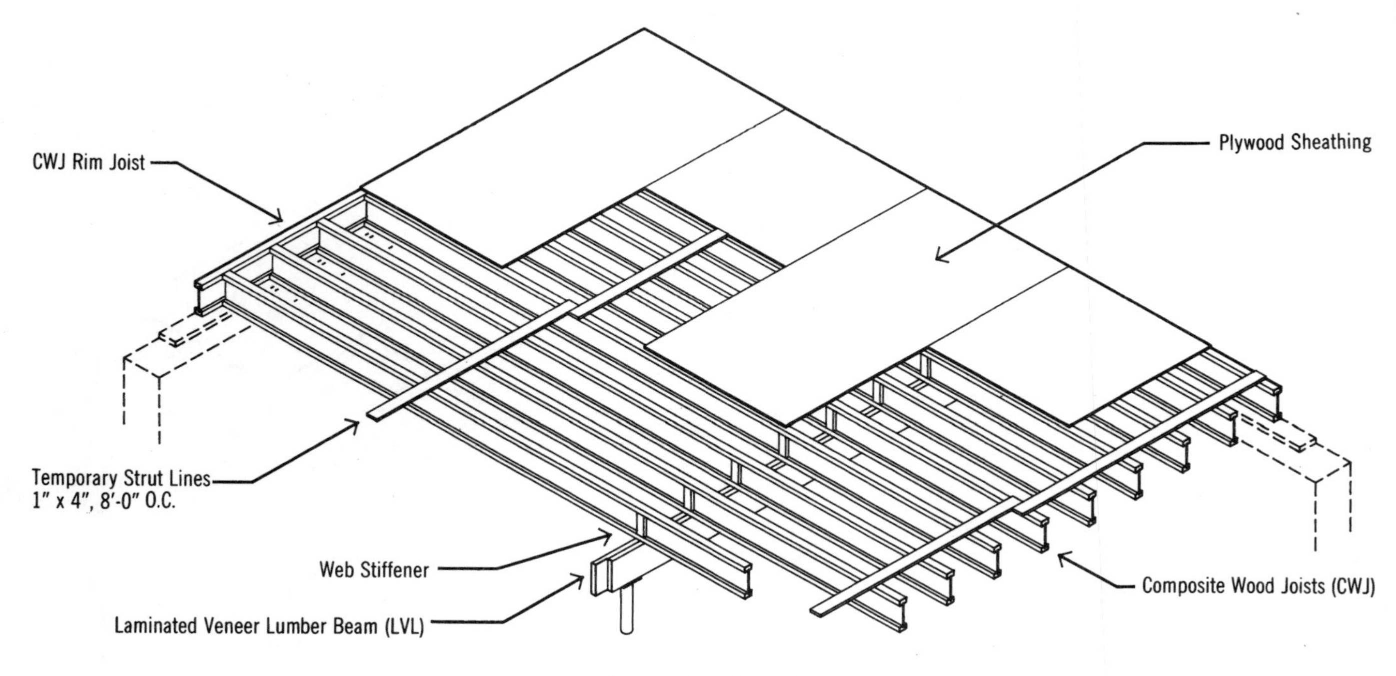

Composite Wood Joists

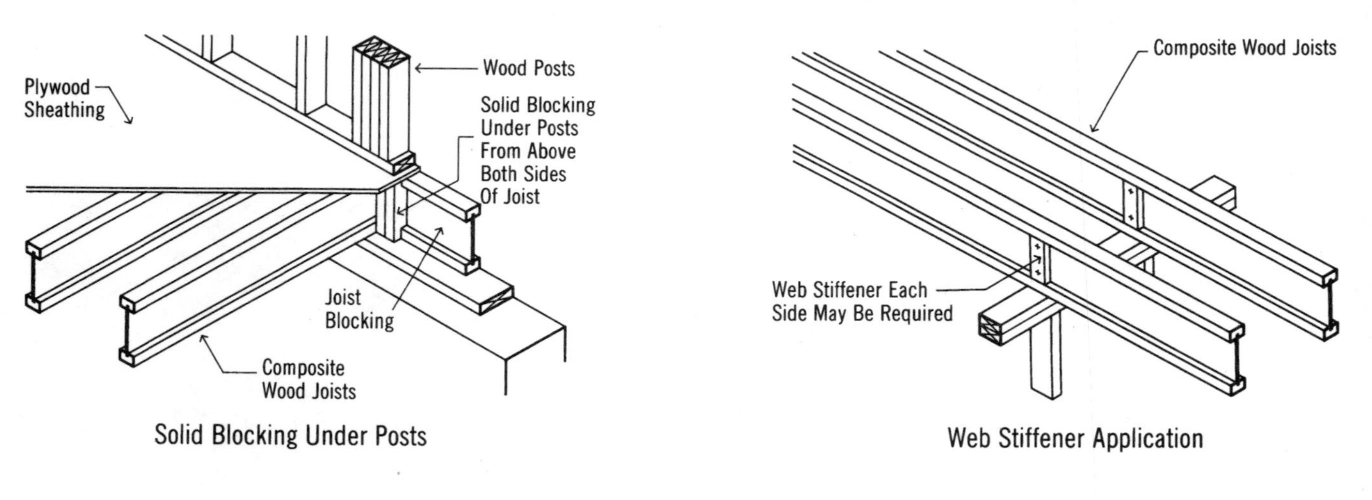

Solid Blocking Under Posts

Web Stiffener Application

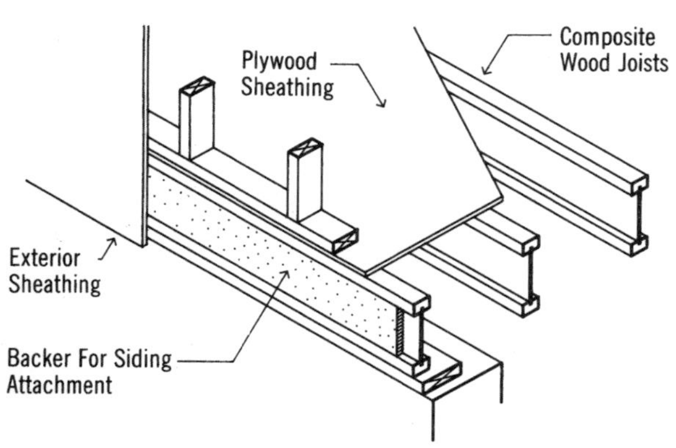

Rim Joist Backer Board Attachment

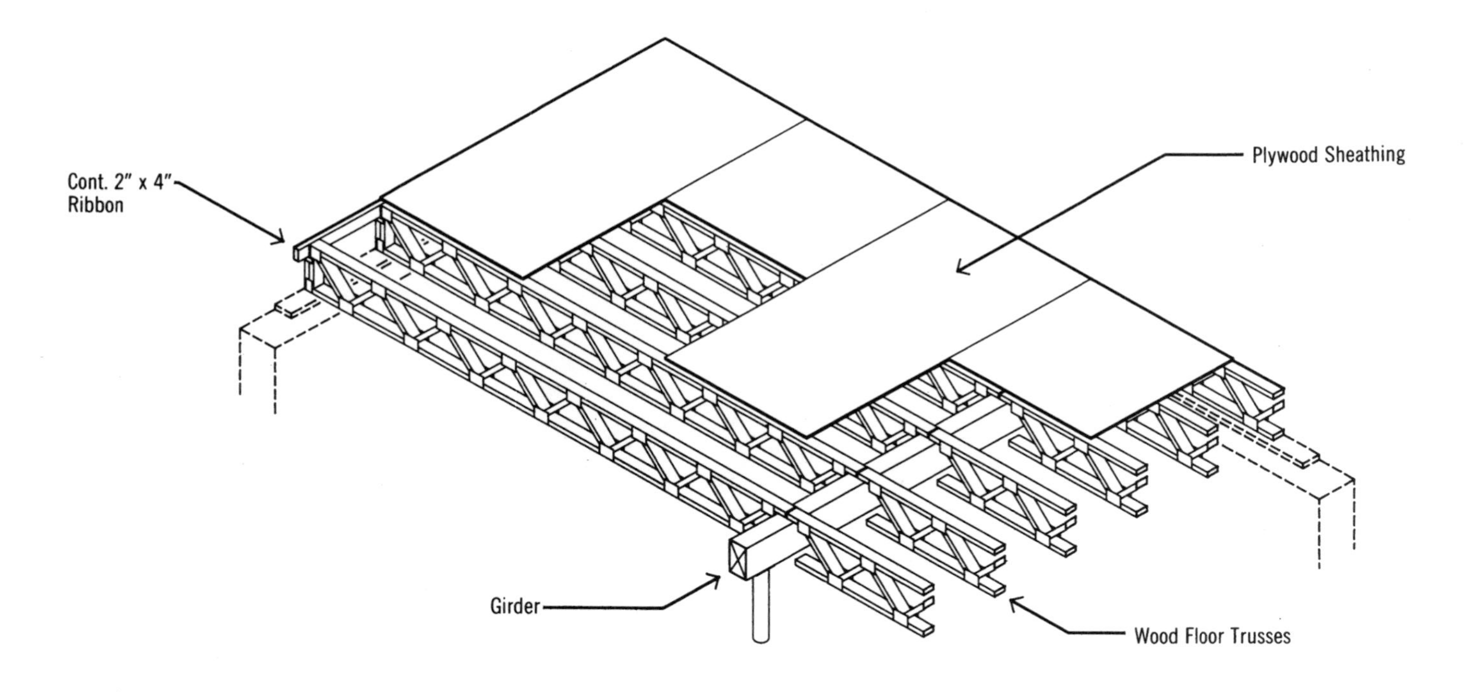

Wood Floor Trusses

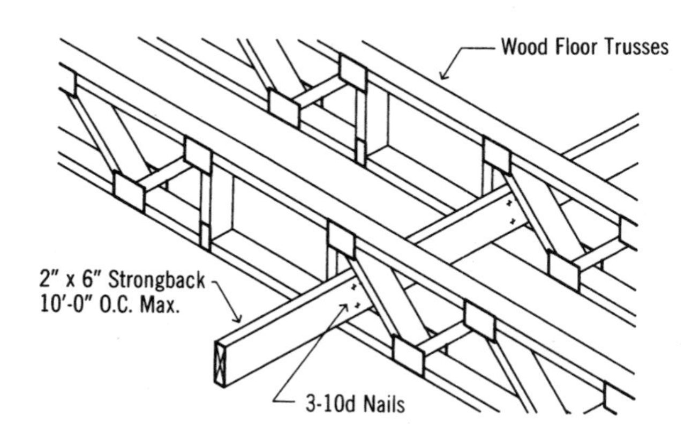

Application of Strongback

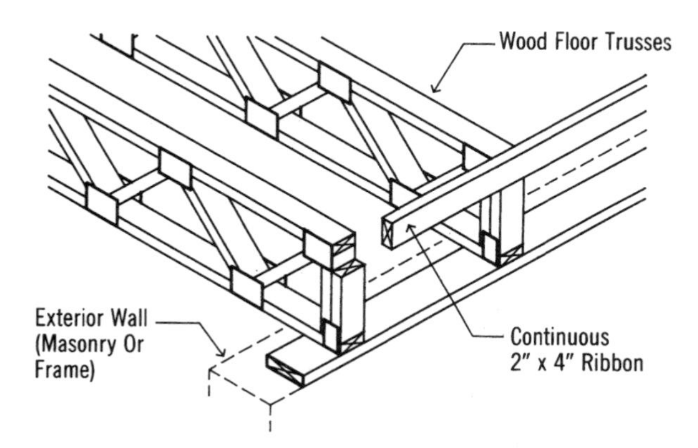

Bottom Chord Exterior Wall Bearing

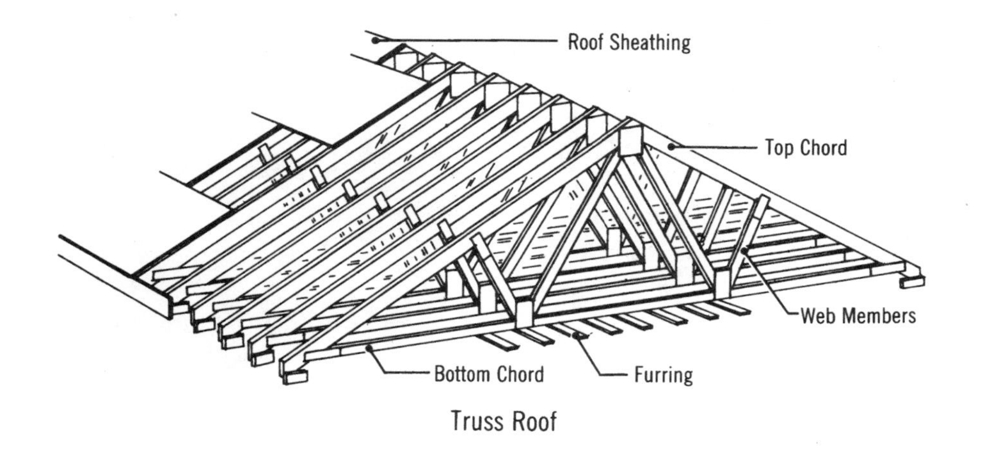

Truss Roof

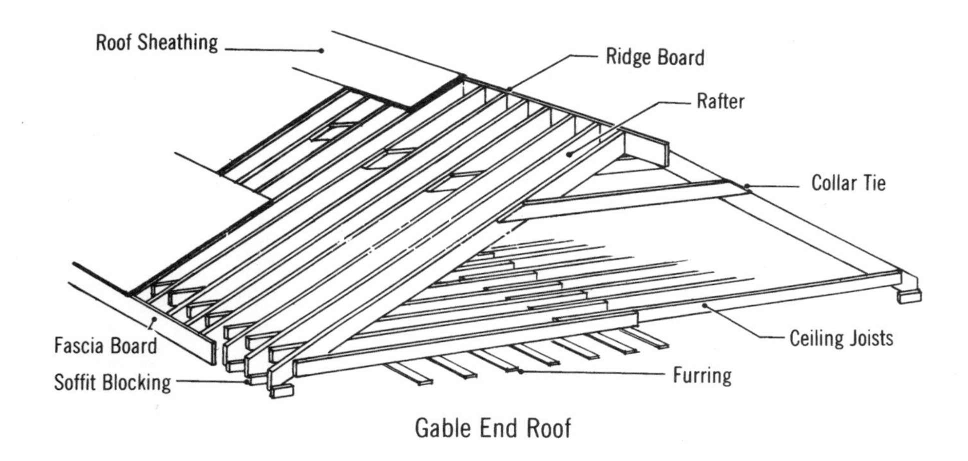

Gable End Roof

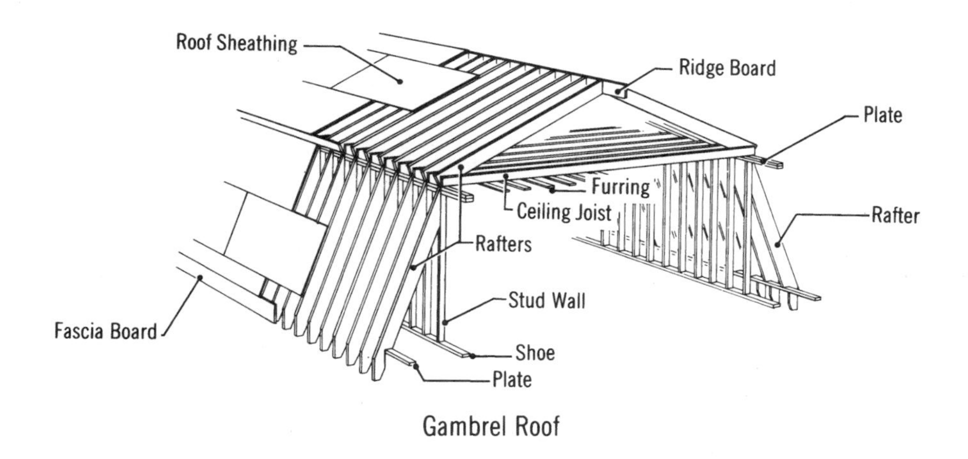

Gambrel Roof

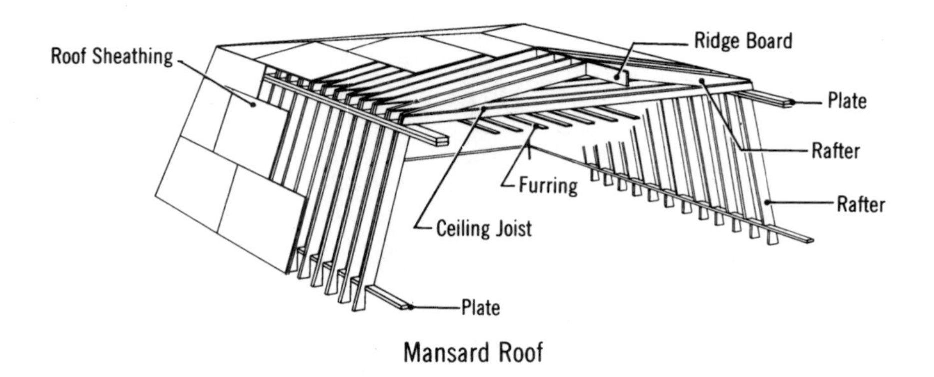

Mansard Roof

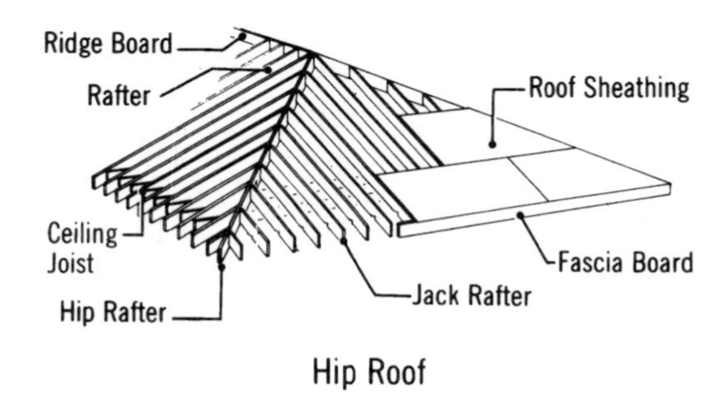

Hip Roof

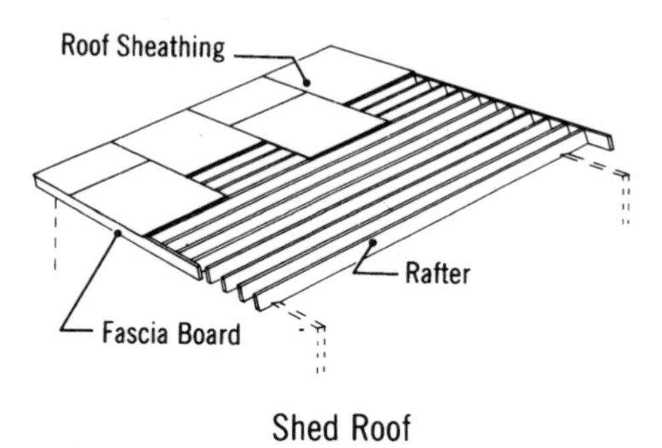

Shed Roof

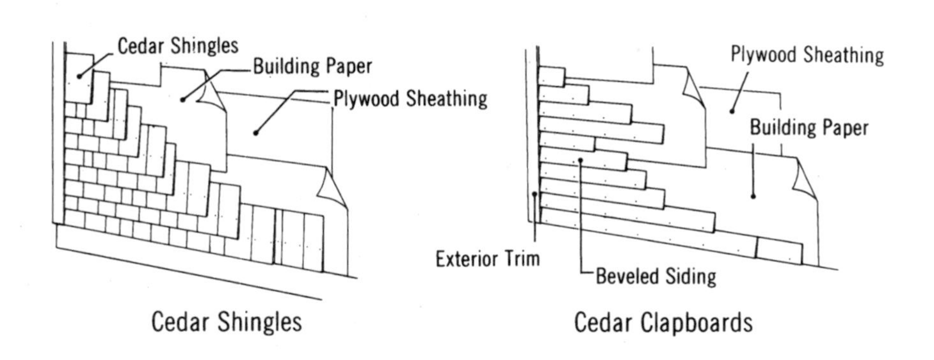

Cedar Shingles

Cedar Clapboards

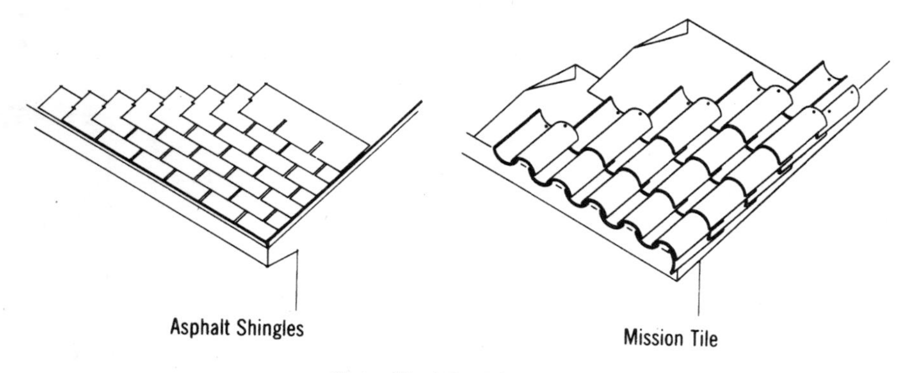

Asphalt Shingles

Mission Tile

Water Shed Roof Coverings

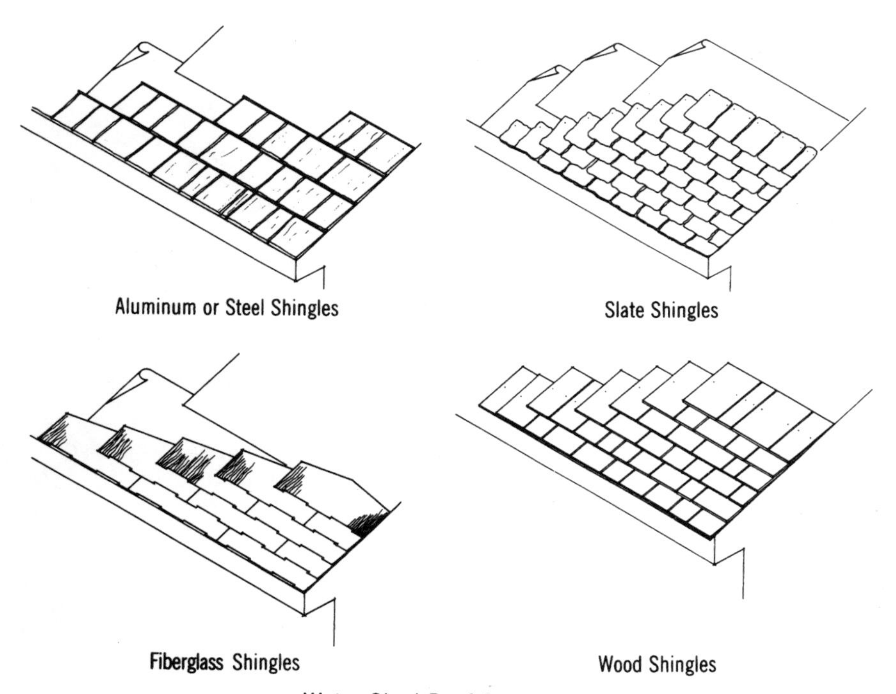

Aluminum or Steel Shingles

Slate Shingles

Fiberglass Shingles

Wood Shingles

Water Shed Roof Coverings

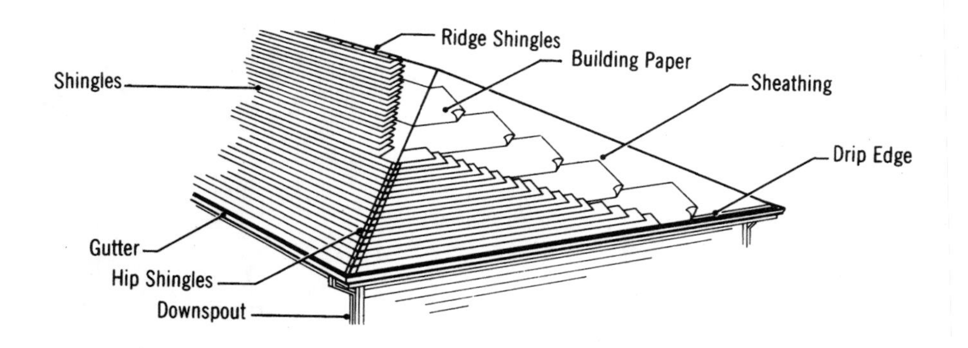

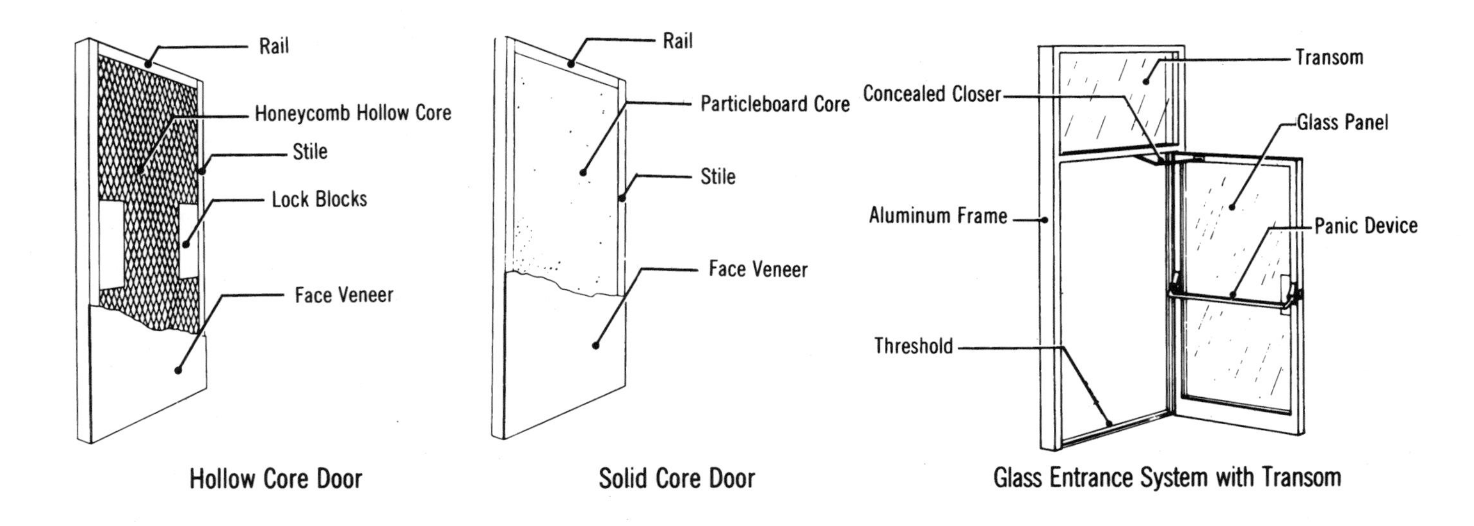

Hollow Core Door

Solid Core Door

Glass Entrance System with Transom

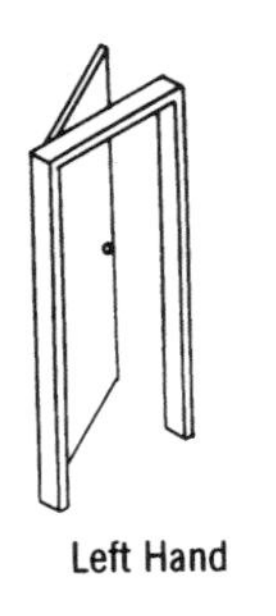

Left Hand

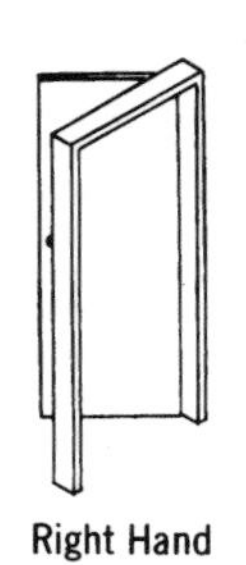

Right Hand

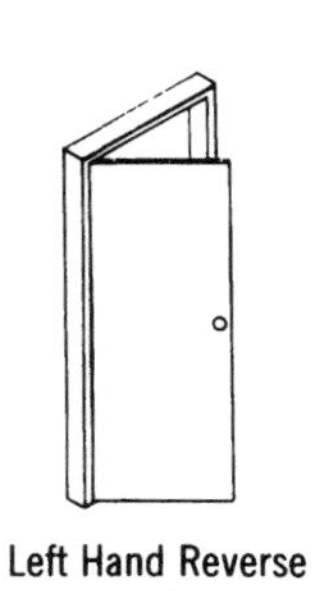

Left Hand Reverse

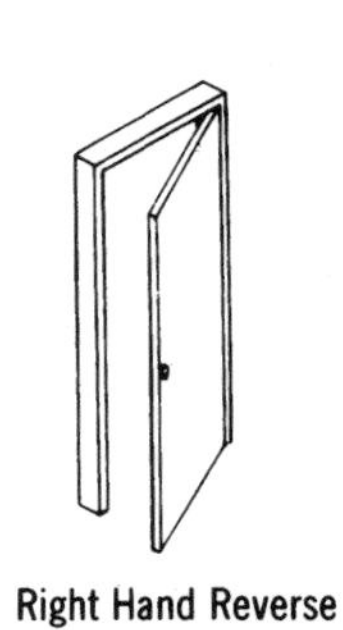

Right Hand Reverse

Hand Designations

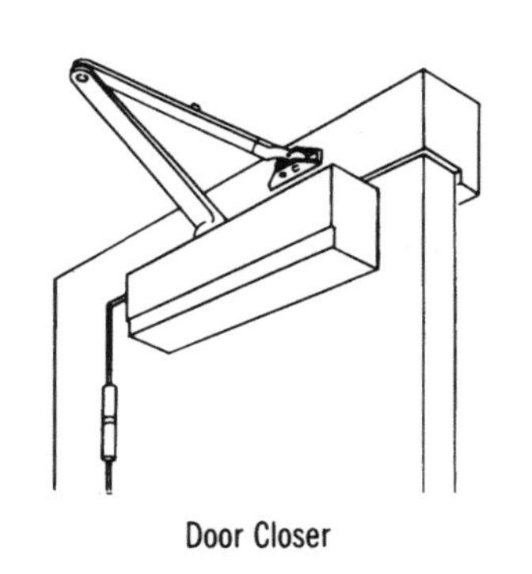

Door Closer

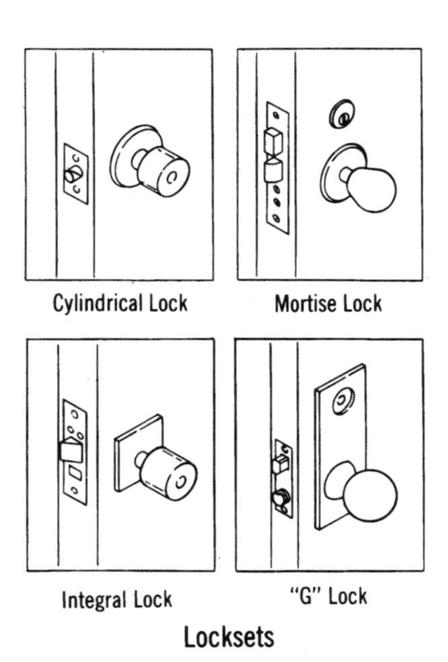

Cylindrical Lock

Mortise Lock

Integral Lock

"G" Lock

Locksets

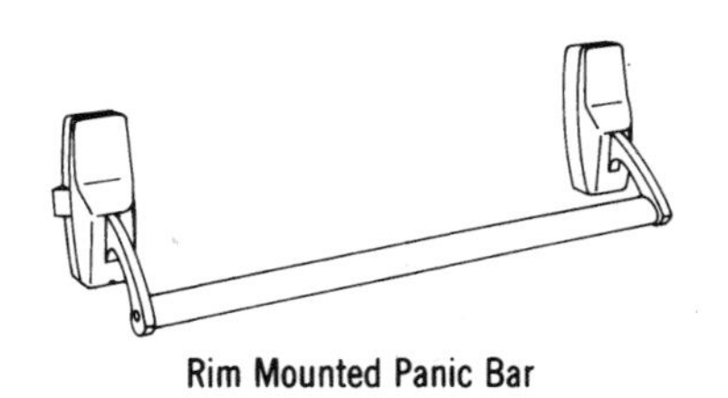

Rim Mounted Panic Bar

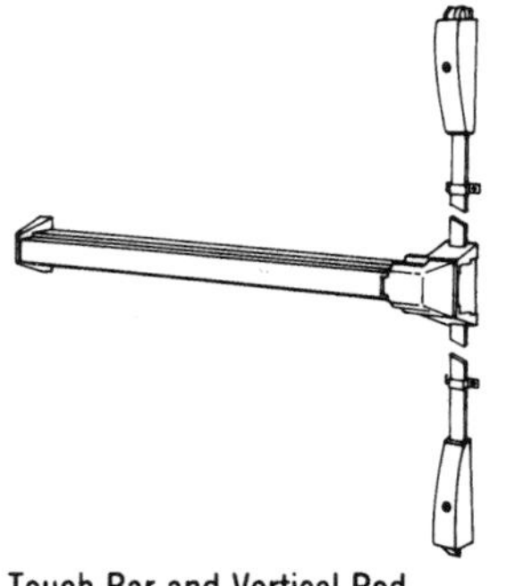

Touch Bar and Vertical Rod

Panic Devices

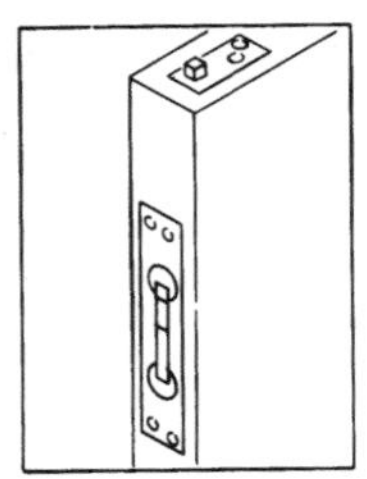

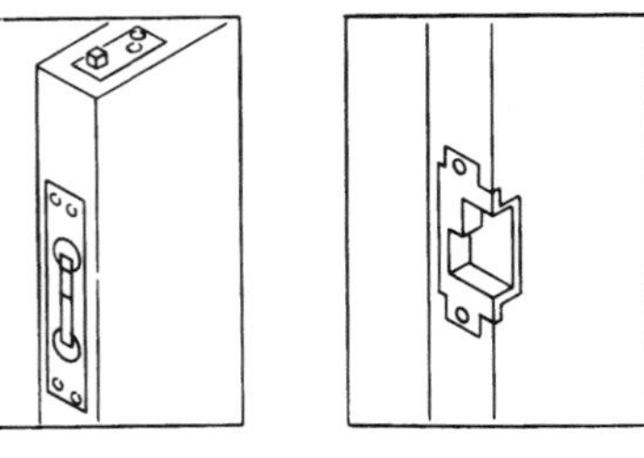

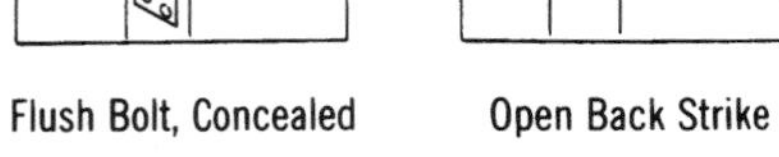

Flush Bolt, Concealed

Open Back Strike

Hardware Nomenclature

Window Terminology

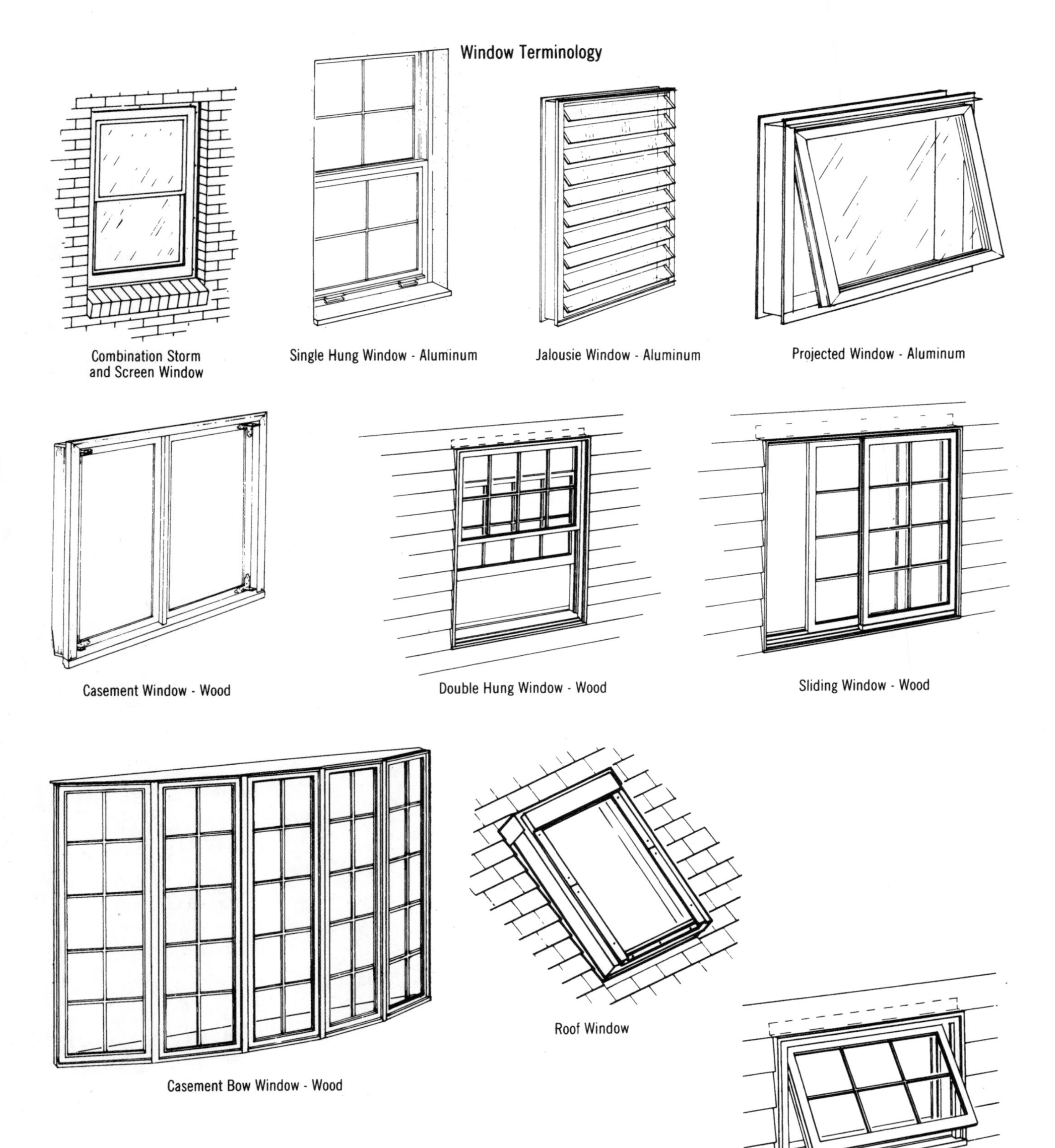

Awning Window - Wood

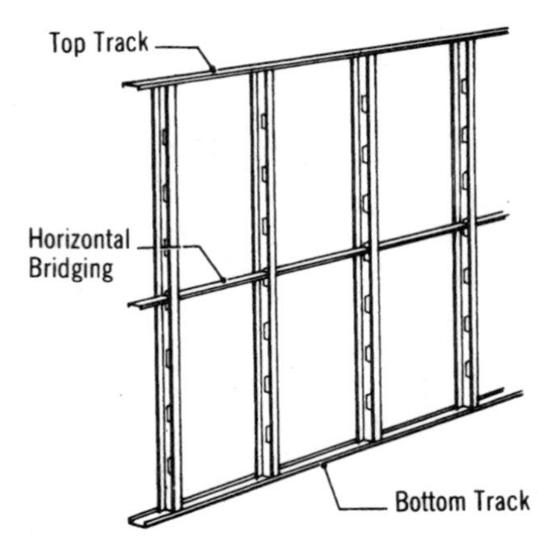

Load Bearing Steel Studs

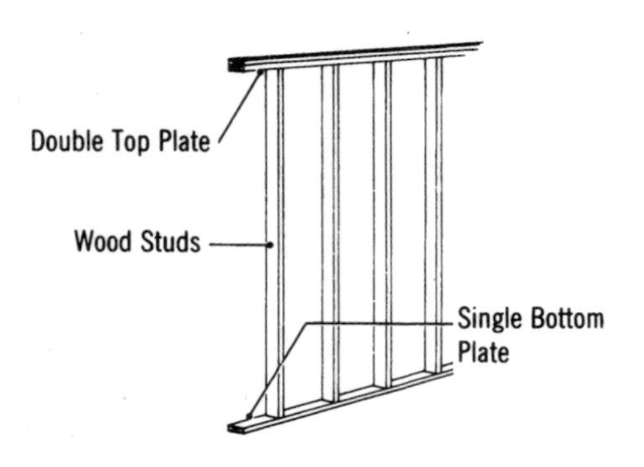

Wood Stud Partition, No Blocking

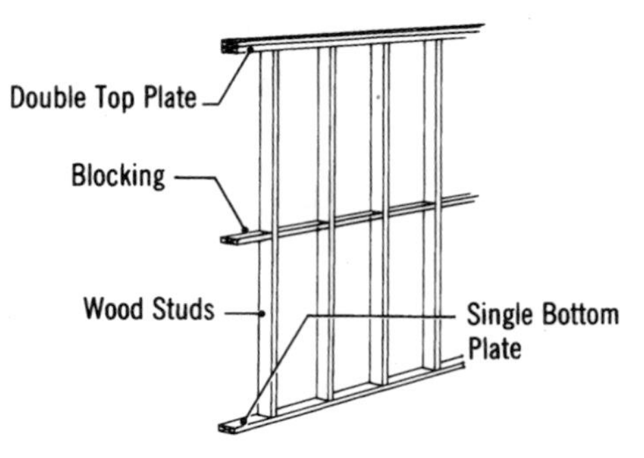

Wood Stud Partition with Blocking

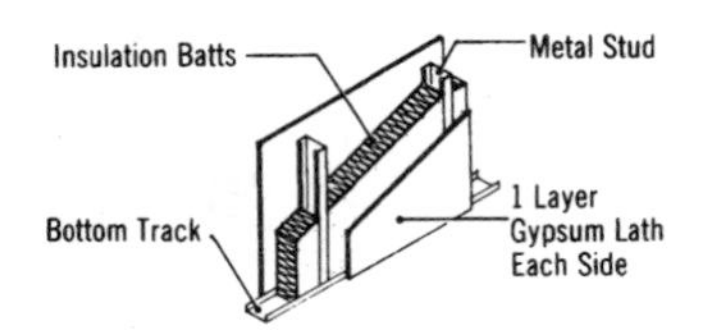

Fiberglass Batt Insulation

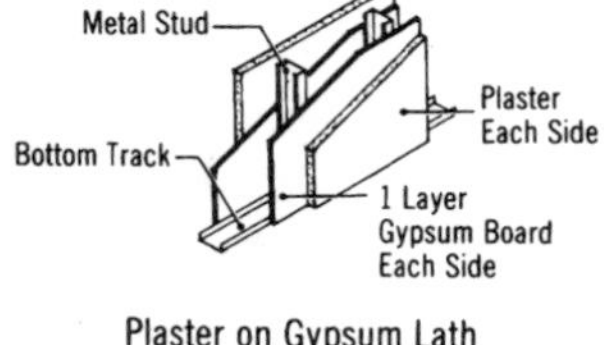

Plaster on Gypsum Lath

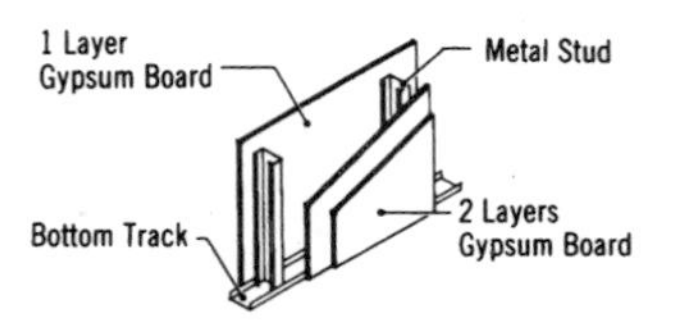

Gypsum Plasterboard, 2 Layers

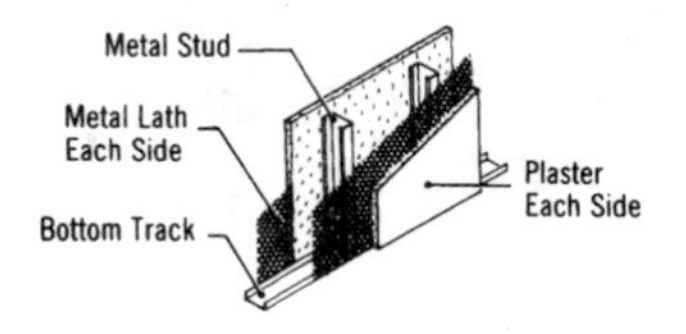

Plaster on Metal Lath

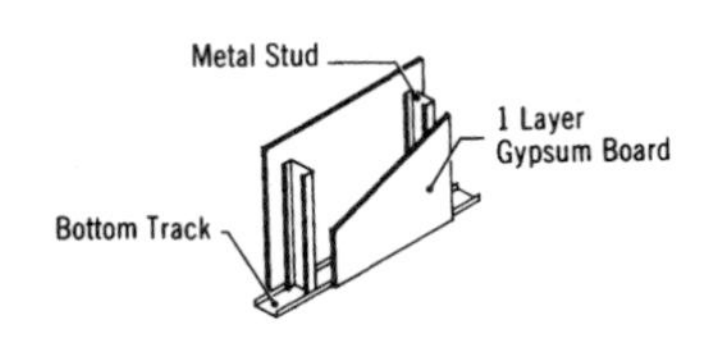

Gypsum Plasterboard

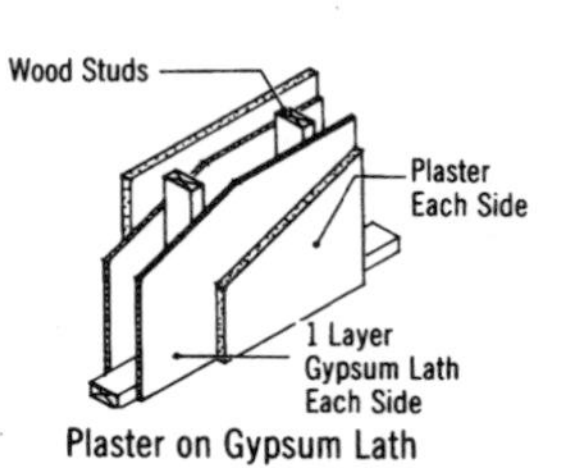

Plaster on Gypsum Lath

Wood Studs
Plaster Each Side
Metal Lath Each Side

Plaster on Metal Lath

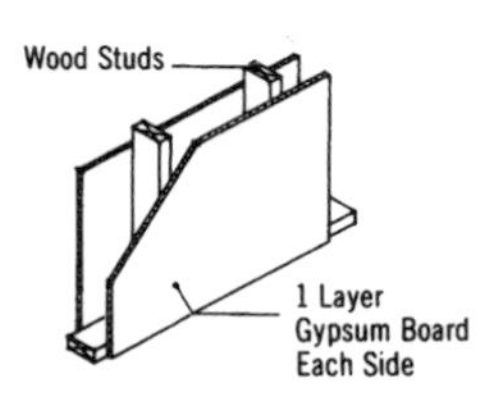

Gypsum Plasterboard

1 Layer Gypsum Board Each Side
Metal Stud
1 Layer Sound-deadening Board Each Side
Bottom Track

Sound-deadening Board

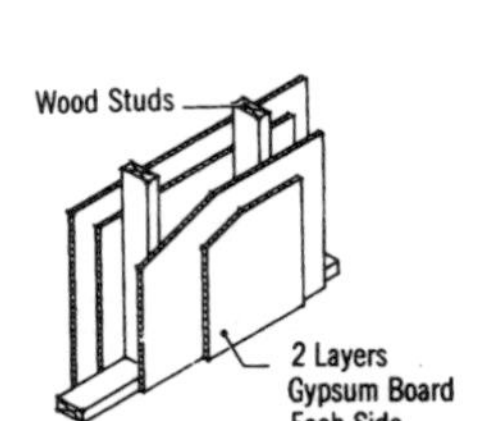

Gypsum Plasterboard, 2 Layers

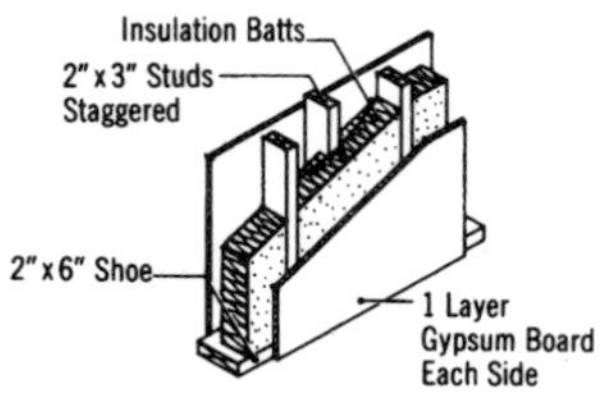

Staggered Stud Wall

Wood Studs
1 Layer Gypsum Board Each Side
1 Layer Sound-deadening Board Each Side

Sound-deadening Board

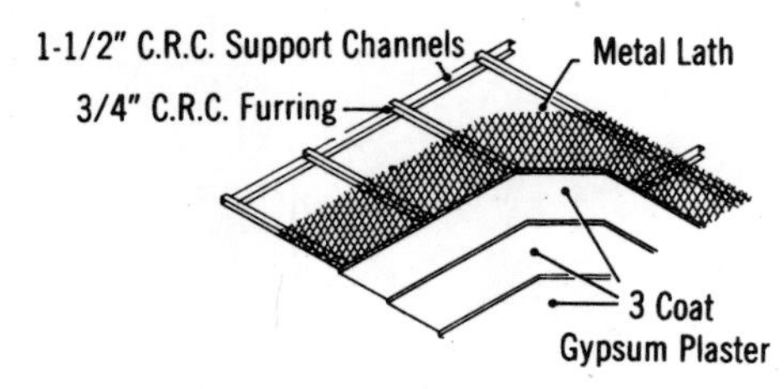

Plaster on Metal Lath

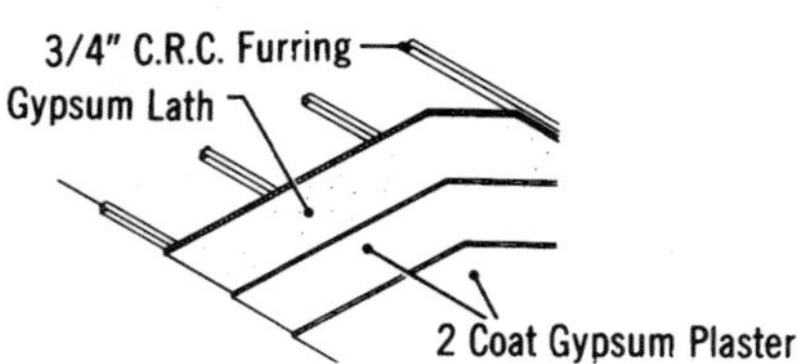

Plaster on Metal Furring

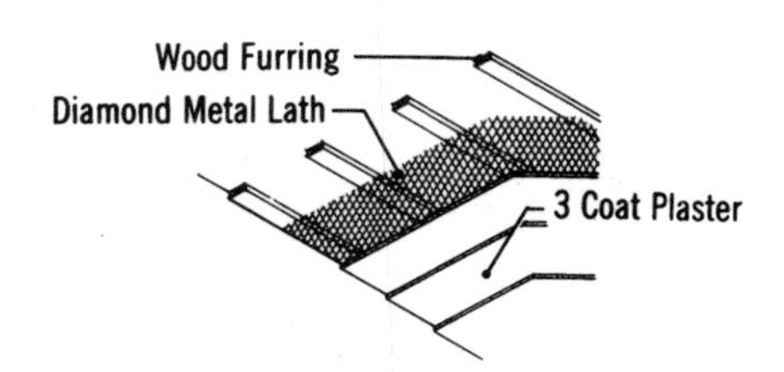

Plaster on Metal Lath and Wood Furring

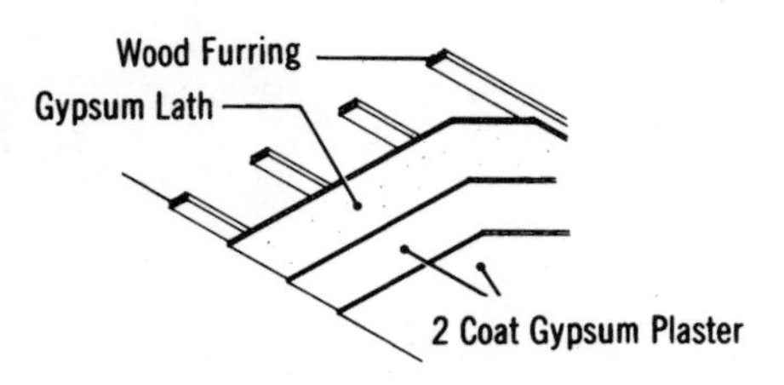

Plaster on Wood Furring

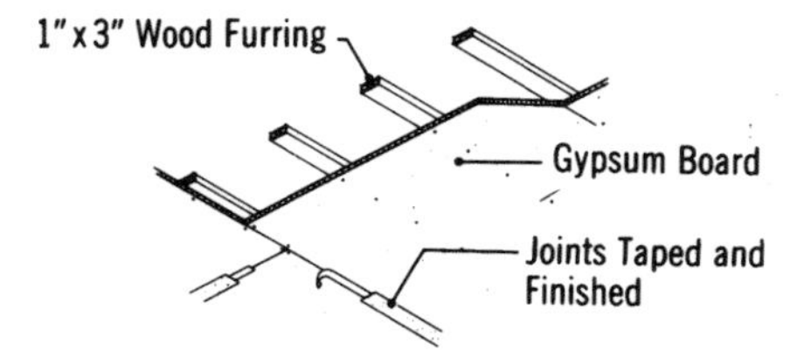

Gypsum Board on 1"x3" Wood Furring

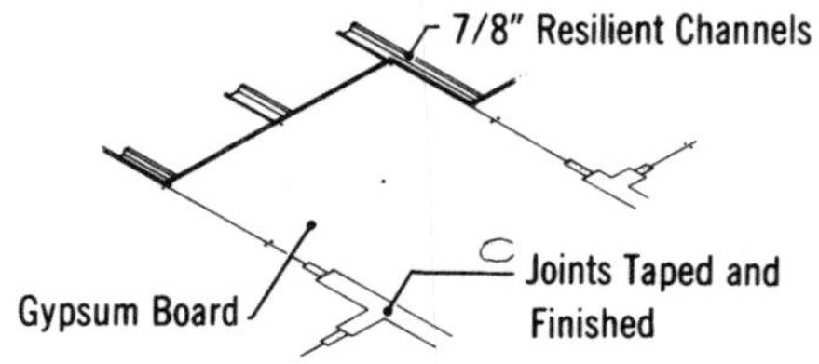

Gypsum Board on 7/8" Resilient Channel Furring

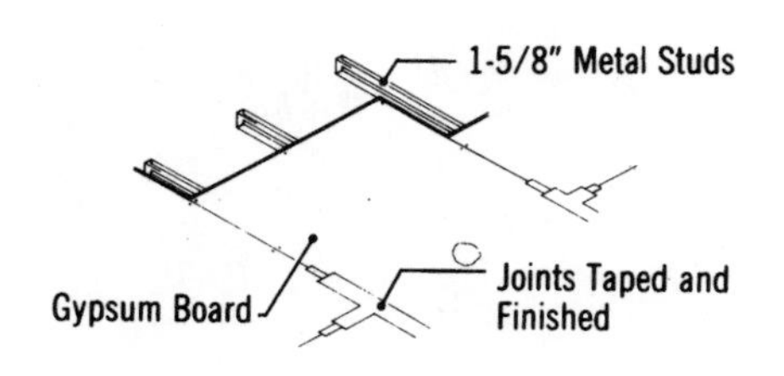

Gypsum Board on 1-5/8" Metal Stud Furring

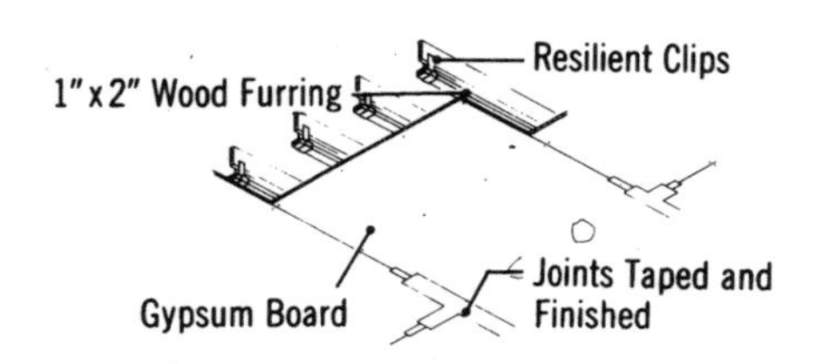

Gypsum Board on 1"x2", Suspended, with Resilient Clips

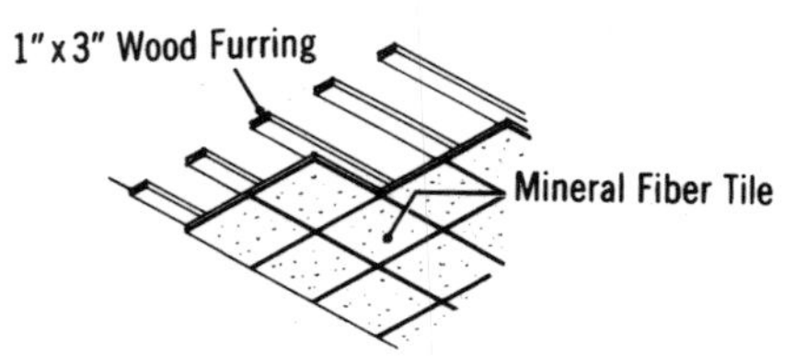

Acoustical Mineral Fiber Tile on 1"x3" Wood Furring

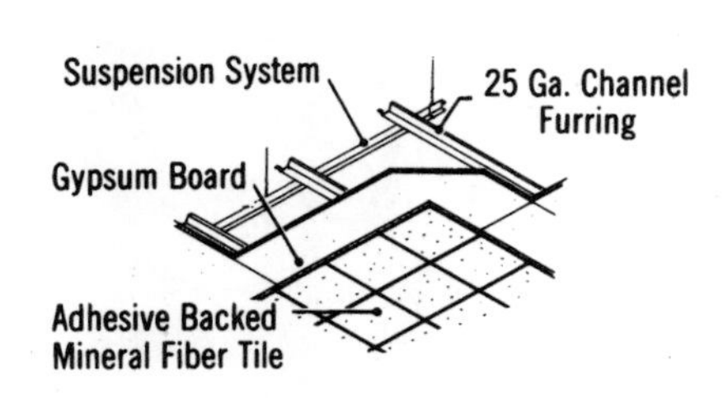

Acoustical Mineral Fiber Tile on Gypsum Board

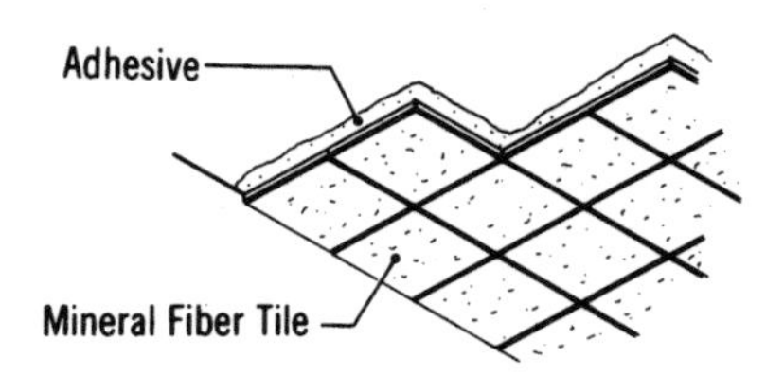

Mineral Fiber Tile Applied with Adhesive

Section Three
Business Information

Total project costs include more than costs for the material, labor, and equipment necessary for the completion of your project. These cost categories include project overhead, general conditions, profit, fees, permits, main office expenses, insurance, tools, taxes, and clean-up to name a few.

This business information section not only provides you with insight into these additional cost factors, but also includes some solid information concerning how to select the most desirable projects for your company to bid, how to manage your existing projects, and some basics concerning scheduling your projects for better profitability.

Business Information

An estimate can be divided into two main groups of costs — direct and indirect. By definition direct costs are those costs that can be directly linked to the project, those costs without which the project could not be completed. Included in this group are material, labor, equipment, and subcontractor costs. These costs are sometimes labeled "bare" or "unburdened" costs.

Direct costs of a construction project include material, labor, equipment (used on the job site), subcontractors, and project overhead.

Material quantities, when taken off carefully, can be turned into highly accurate material costs. For this to happen your costs must be current and reliable. The most reliable source for the information is a quote from a reputable vendor for that project. A material supplier should not be chosen on price alone. You must ask if the vendor can deliver the materials at the price quoted. Will they add "special charges" for shipping, delivery, or unloading? Will they be on time? Why are they less expensive? Are there any time limits or escalation clauses on the price? Is this price for the specified materials or is this for an "or equal"? Is this cash up front, C.O.D., or on credit? Is this a stock item that is readily available or will it have to be special ordered? How long will it take? What type of guarantee will they give? Many people would rather pay a higher price to obtain materials from a known, reliable source than risk the unknown. However, there are plenty of bargains out there if you are willing to take some risks.

Other sources of cost information include manufacturers' catalogs (be sure to call and get the contractors' prices or the discount), current cost records from previous jobs, and reputable cost guides.

In order to determine the **labor** costs you need to have three pieces of information: the number of units to be installed, the productivity (how long it takes to install a unit), and the hourly wage. The number of units to be installed are obtained from the material quantity. The productivity of a worker or a crew is very difficult to determine. If you are familiar with a worker or a crew that has been working together for a while, you will have a general "feel" for how much they can install in a day's or week's time. Therefore, the best place to get productivity factors is from your own accurate records from previous projects.

Wage rates are not difficult to find, in most cases. If the project is a union project, then wage rates will be available to you from the union locals. If you are doing an estimate for your own company then you should be able to get the rates from your company's files. You may run into a problem if you are estimating an open shop or non-union project in an unfamiliar location. Often there are organizations in or near larger cities that can assist you in locating this information. Lacking such help, you will have to do some old fashioned "detective work" to get the rates you need.

Equipment costs generally do not include the cost of the operators. In cases where the equipment is listed or leased as an "operated" piece of equipment, the price includes the operator. The costs for rented or leased pieces of equipment can be obtained from local dealers, suppliers, or even manufacturers. These costs can and will fluctuate and need to be updated regularly. Whether you rent or own you will need to add into the overall costs any operating costs: the costs for fuel and lubrication, maintenance (if owned or if in the lease agreement), parts, transportation costs, and mobilization costs. Normally, manufacturers and leasing agents should be able to provide you with these costs.

You can include the equipment costs in two ways. The first is to include the costs as part of the job task; the second is to consider the equipment as part of the overhead of the whole project. The advantage of putting the equipment in with the task is that when reviewing the job records for a future project you will have the equipment costs readily available and thus you will get a clearer picture of the true job cost. Whichever method you choose, be consistent. Make sure that all equipment costs are included and are not duplicated.

Subcontractor quotes should be analyzed in the same fashion as material quotes and the same questions apply. Make sure the subcontractor quote covers **all** the plans and specifications as required. Any exclusions, allowances, alternates, time limits to the bid, and time allowances for work should be duly noted and clearly explained. Keep in mind that you may not look at the estimate until months later. If any subcontractor bids are received by phone, immediately confirm the bid and **all** the details in **writing**. Verbal agreements are subject to a great deal of misunderstanding and misinterpretation.

Project overhead needs to be calculated for this project. Project overhead can be defined as the sum of those items that are necessary for the actual construction of the project, but are not identifiable with any particular work item. The checklist provided in this section specifies some of the items that need to be included. You may not agree with all the items on the list because in some cases you may be able to associate some items with a direct work item. It is not important **where** you place these items as long as you do include them in your estimate.

All the direct costs should be itemized, priced, tabulated, and totaled before the indirect costs are added. Indirect costs are usually added to an estimate at the summary stage and are, in many cases, calculated as a percentage of the direct costs. Included in this grouping are such items as sales tax on materials, overhead, profit, and contingencies. These indirect costs generally account for the variations among estimates.

Overhead refers to all costs involved in operating your business. It is essential to your company's very existence that these costs be factored into and proportioned over all your projects. The following checklist in this section indicates the expenses that should be covered. The following example indicates how overhead percentages can be computed.

A-Z-Z Contracting, Anytown, USA
Annual Volume = $1,000,000.
Home Office (indirect) Overhead
Calculation — Annual Costs

Salaries

Owner	$40,000
Secretary	15,000
Bookkeeper (part time)	9,000
Accountant (retainer)	4,000
Legal fees (retainer)	4,000
Workers' Comp./medical	7,000
Office	
Rent	8,000
Equipment & utilities	7,000
Supplies	2,000
Advertising	3,000
Auto/Truck	5,000
Association fees	4,000
Entertainment	2,000
Bad debt	5,000
Total	$115,000

Therefore, for each dollar in an estimate, this contractor must add $115,000/$1,000,000 or 11.5 percent of the total for office overhead.

The following charts, checklists, and data are to be used to help remind you of what items should be included, and their common ranges, in calculating your direct and indirect expenses.

Estimate Summary

After the estimate has been reviewed and all changes have been incorporated, the total for each phase of the work needs to be summarized. Except for estimates with a limited number of items, it is recommended that the costs be transferred from the estimating sheets to an estimate summary sheet. The transferring of the costs should be double-checked, as it is easy to transpose numbers and/or make other simple errors that could cost you thousands of dollars.

The estimate summary sheet should contain the following columns:

Material
Labor
Equipment
Subcontractor
Totals

Listing items in their proper columns makes it mathematically simple to arrive at the sum for each category and to apply the appropriate markups to the total dollar values for each column. Each category would normally have different markup percentages added at the end of the estimate for:

Insurance, taxes, fringe benefits
Equipment and tools (if equipment is being treated as overhead items)
General overhead, profit, contingencies, and fees

The final total on the estimate sheet is the total estimate. If this number is used as the bid figure, you will be committed to it upon acceptance. It is, therefore, very worthwhile to take the time to check that this figure is correct.

A Good Bid

As a bidder, your immediate goal is to submit a bid which is as low as possible but allows you to make a reasonable profit on the job. Therefore you want your bid to be a **good bid**. To define a **good** bid, let's look at the **ideal** bid.

The ideal bid is one that is submitted on work that would be great to get but not a financial loss if you don't, is fair to all parties, allows you to make a good profit, and involves no risk to you. Such a bid is a rarity. In the real world the best we can hope for would be a **good** bid.

A good bid will have the following qualities:

a. be low but not out of line with other bids; in other words, very competitive.
b. involve reasonable to low risk.
c. contain accurate quantities.
d. contain accurate prices from reputable suppliers and subcontractors.
e. provide completeness of coverage.
f. set realistic deadlines.
g. describe realistic plans for execution.
h. include appropriate markups.
i. ensure profitability.
j. utilize properly executed bid forms.

As with any skill, your ability to produce accurate and competitive bids will improve with experience. You can reasonably expect your bidding success rate to improve as your experience grows. Your goal is always to bid low — and show a profit.

Checklist

For an estimate to be reliable, all items must be accounted for. A complete estimate can also eliminate the need to include contingencies. The following checklist can be used to help ensure that all items are properly accounted for.

Direct Overhead Costs

Personnel

- ☐ Superintendent
- ☐ Project Manager (if for that project only)
- ☐ Field Engineer (if for that project only)
- ☐ Cost Engineer (if for that project only)
- ☐ Warehouse personnel (if for that project only)
- ☐ Watchman/guard dogs
- ☐ Tool room keeper (if for that job only)
- ☐ Timekeeper (if for that job only)
- ☐ Foreman (working directly for the contractor)

Temporary Facilities

- ☐ Field office expense
  - ____ Set-up and removal
  - ____ Light
  - ____ Heat
  - ____ Water
  - ____ Telephone
  - ____ Supplies
  - ____ Equipment
    - ____ Fax machine
    - ____ Copy machine
    - ____ Blueprint machine
    - ____ Coffee machine
- ☐ Temporary light and power
- ☐ Temporary heat
- ☐ Temporary water
- ☐ Pay telephones
- ☐ Toilet facilities
- ☐ Enclosures
- ☐ Storage trailers
- ☐ Fencing
- ☐ Barricades and signals
- ☐ Construction road
- ☐ Job sign

Miscellaneous

- ☐ Vehicles
- ☐ Permits
- ☐ Licenses
- ☐ Tools and equipment
- ☐ Photographs
- ☐ Surveying
- ☐ Testing
- ☐ Job signs
- ☐ Pumping
- ☐ Dust control
- ☐ Lifting/hoisting
- ☐ Cleanup (periodical)
- ☐ Final cleanup
- ☐ Damage/repair to adjoining buildings and/or public ways

Indirect Costs

Salaries

- ☐ President
- ☐ Executives
- ☐ Secretaries/Reception
- ☐ Estimators
- ☐ Project Managers
- ☐ Construction Manager
- ☐ Cost Engineers
- ☐ Purchasing Agent
- ☐ Cost/Bookkeeping
- ☐ Engineers
- ☐ Other office personnel
- ☐ Yard personnel
  - ____ Tool Manager
  - ____ Mechanics/Maintenance
  - ____ Drivers
  - ____ Equipment operators

Office

- ☐ Rent/cost of ownership
- ☐ Electricity
- ☐ Gas
- ☐ Water
- ☐ Sewer
- ☐ Telephone
- ☐ Postage
- ☐ Office equipment
- ☐ Furniture/furnishings
- ☐ Office supplies
- ☐ Advertising
- ☐ Literature
- ☐ Club/association dues

Professional Services

- ☐ Legal
- ☐ Accounting
- ☐ Architectural
- ☐ Engineering

Vehicles

- ☐ Cars/trucks
- ☐ Cost of operation
- ☐ Mileage expenses

Insurance

- ☐ Fire
- ☐ Property damage
- ☐ Vehicles
- ☐ Public liability
- ☐ Windstorm
- ☐ Workers' Compensation
- ☐ Unemployment
- ☐ Social Security
- ☐ Flood
- ☐ Theft
- ☐ Elevator

Bonds

- ☐ Bid
- ☐ Payment
- ☐ Performance
- ☐ Surety
- ☐ Lien

Installing Contractor's Overhead & Profit

Below are the **average** installing contractor's percentage mark-ups applied to base labor rates to arrive at typical billing rates.

Column A: Labor rates are based on average open shop wages for 7 major U.S. regions. Base rates including fringe benefits are listed hourly and daily. These figures are the sum of the wage rate and employer-paid fringe benefits such as vacation pay, and employer-paid health costs.

Column B: Workers' Compensation rates are the national average of state rates established for each trade.

Column C: Column C lists average fixed overhead figures for all trades. Included are Federal and State Unemployment costs set at 7.0%; Social Security Taxes (FICA) set at 7.65%; Builder's Risk Insurance costs set at 0.34%; and Public Liability costs set at 1.55%. All the percentages except those for Social Security Taxes vary from state to state as well as from company to company.

Columns D and E: Percentages in Columns D and E are based on the presumption that the installing contractor has annual billing of $1,000,000 and up. Overhead percentages may increase with smaller annual billing. The overhead percentages for any given contractor may vary greatly and depend on a number of factors, such as the contractor's annual volume, engineering and logistical support costs, and staff requirements. The figures for overhead and profit will also vary depending on the type of job, the job location, and the prevailing economic conditions. All factors should be examined very carefully for each job.

Column F: Column F lists the total of Columns B, C, D, and E.

Column G: Column G is Column A (hourly base labor rate) multiplied by the percentage in Column F (O&P percentage).

Column H: Column H is the total of Column A (hourly base labor rate) plus Column G (Total O&P).

Column I: Column I is Column H multiplied by eight hours.

		A		B	C	D	E	F	G	H	I
		Base Rate Incl. Fringes		Work-ers' Comp. Ins.	Average Fixed Over-head	Over-head	Profit	Total Overhead & Profit		Rate with O & P	
Abbr.	Trade	Hourly	Daily					%	Amount	Hourly	Daily
Skwk	Skilled Workers Average (35 trades)	$20.50	$164.00	17.5%	16.5%	27.0%	10%	71.0%	$14.55	$35.05	$280.40
	Helpers Average (5 trades)	15.25	122.00	19.3		25.0		70.8	10.80	26.05	208.40
	Foreman Average, Inside ($.50 over trade)	21.00	168.00	17.5		27.0		71.0	14.90	35.90	287.20
	Foreman Average, Outside ($2.00 over trade)	22.50	180.00	17.5		27.0		71.0	16.00	38.50	308.00
Clab	Common Building Laborers	14.85	118.80	19.0		25.0		70.5	10.45	25.30	202.40
Asbe	Asbestos/Insulation Workers/Pipe Coverers	21.20	169.60	17.7		30.0		74.2	15.75	36.95	295.60
Boil	Boilermakers	22.80	182.40	15.9		30.0		72.4	16.50	39.30	314.40
Bric	Bricklayers	20.70	165.60	17.0		25.0		68.5	14.20	34.90	279.20
Brhe	Bricklayer Helpers	16.05	128.40	17.0		25.0		68.5	11.00	27.05	216.40
Carp	Carpenters	20.40	163.20	19.0		25.0		70.5	14.40	34.80	278.40
Cefi	Cement Finishers	19.55	156.40	11.1		25.0		62.6	12.25	31.80	254.40
Elec	Electricians	23.00	184.00	6.8		30.0		63.3	14.55	37.55	300.40
Elev	Elevator Constructors	23.90	191.20	8.4		30.0		64.9	15.50	39.40	315.20
Eqhv	Equipment Operators, Crane or Shovel	21.50	172.00	10.9		28.0		65.4	14.05	35.55	284.40
Eqmd	Equipment Operators, Medium Equipment	20.95	167.60	10.9		28.0		65.4	13.70	34.65	277.20
Eqlt	Equipment Operators, Light Equipment	19.85	158.80	10.9		28.0		65.4	13.00	32.85	262.80
Eqol	Equipment Operators, Oilers	17.75	142.00	10.9		28.0		65.4	11.60	29.35	234.80
Eqmm	Equipment Operators, Master Mechanics	21.95	175.60	10.9		28.0		65.4	14.35	36.30	290.40
Glaz	Glaziers	20.25	162.00	14.1		25.0		65.6	13.30	33.55	268.40
Lath	Lathers	19.75	158.00	11.8		25.0		63.3	12.50	32.25	258.00
Marb	Marble Setters	20.45	163.60	17.0		25.0		68.5	14.00	34.45	275.60
Mill	Millwrights	21.40	171.20	11.2		25.0		62.7	13.40	34.80	278.40
Mstz	Mosaic & Terrazzo Workers	19.80	158.40	10.3		25.0		61.8	12.25	32.05	256.40
Pord	Painters, Ordinary	18.55	148.40	14.8		25.0		66.3	12.30	30.85	246.80
Psst	Painters, Structural Steel	19.30	154.40	50.0		25.0		101.5	19.60	38.90	311.20
Pape	Paper Hangers	18.65	149.20	14.8		25.0		66.3	12.35	31.00	248.00
Pile	Pile Drivers	20.15	161.20	27.7		30.0		84.2	16.95	37.10	296.80
Plas	Plasterers	19.10	152.80	15.7		25.0		67.2	12.85	31.95	255.60
Plah	Plasterer Helpers	16.30	130.40	15.7		25.0		67.2	10.95	27.25	218.00
Plum	Plumbers	22.75	182.00	8.5		30.0		65.0	14.80	37.55	300.40
Rodm	Rodmen (Reinforcing)	21.95	175.60	29.2		28.0		83.7	18.35	40.30	322.40
Rofc	Roofers, Composition	17.75	142.00	34.3		25.0		85.8	15.25	33.00	264.00
Rots	Roofers, Tile & Slate	17.95	143.60	34.3		25.0		85.8	15.40	33.35	266.80
Rohe	Roofers, Helpers (Composition)	13.20	105.60	34.3		25.0		85.8	11.35	24.55	196.40
Shee	Sheet Metal Workers	22.15	177.20	11.9		30.0		68.4	15.15	37.30	298.40
Spri	Sprinkler Installers	22.75	182.00	8.8		30.0		65.3	14.85	37.60	300.80
Stpi	Steamfitters or Pipefitters	23.00	184.00	8.5		30.0		65.0	14.95	37.95	303.60
Ston	Stone Masons	20.25	162.00	17.0		25.0		68.5	13.85	34.10	272.80
Sswk	Structural Steel Workers	22.00	176.00	40.9		28.0		95.4	21.00	43.00	344.00
Tilf	Tile Layers	19.70	157.60	10.3		25.0		61.8	12.15	31.85	254.80
Tilh	Tile Layers Helpers	15.90	127.20	10.3		25.0		61.8	9.85	25.75	206.00
Trlt	Truck Drivers, Light	16.60	132.80	15.3		25.0		66.8	11.10	27.70	221.60
Trhv	Truck Drivers, Heavy	17.00	136.00	15.3		25.0		66.8	11.35	28.35	226.80
Sswl	Welders, Structural Steel	22.00	176.00	40.9		28.0		95.4	21.00	43.00	344.00
Wrck	*Wrecking	15.30	122.40	42.2		25.0		93.7	14.35	29.65	237.20

*Not included in averages

General Contractor's Overhead

The table below shows a contractor's overhead as a percentage of direct cost in two ways. The figures on the right are for the overhead, markup based on both material and labor. The figures on the left are based on the entire overhead applied only to the labor. This figure would be used if the owner supplied the materials or if a contract is for labor only.

Items of General Contractor's Indirect Costs	% of Direct Costs	
	As a Markup of Labor Only	As a Markup of Both Material and Labor
Field Supervision	6.0%	2.9%
Main Office Expense (see details below)	16.2	7.7
Tools and Minor Equipment	1.0	0.5
Workers' Compensation & Employers' Liability.	17.5	8.3
Field Office, Sheds, Photos, Etc.	1.5	0.7
Performance and Payment Bond, 0.7% to 1.5%.	2.3	1.1
Unemployment Tax	7.0	3.3
Social Security and Medicare.	7.7	3.7
Sales Tax — add if applicable 42/80 x % as markup of total direct costs including both material and labor.		
Sub Total	59.2%	28.2%
*Builder's Risk Insurance ranges from 0.141% to 0.586%.	0.6	0.3
*Public Liability Insurance	3.2	1.5
Grand Total	63.0%	30.0%

*Paid by Owner or Contractor

General Contractor's Main Office Expense for Smaller Projects

This table provides average main office expenses (as a percentage of annual volume) for contractors specializing in smaller projects, such as repair and/or remodeling of existing structures.

Annual Volume	% of Annual Volume	
	Minimum	Maximum
To $50,000	20%	30%
To $100,000	17%	22%
To $250,000	16%	19%
To $500,000	14%	16%
To $1,000,000	8%	10%

Overhead and Profit as a Percentage of Project Costs

Overhead is defined as costs that are associated with a construction project, but not directly with the actual construction. This table shows percentages that can be used as a rule of thumb for estimating overhead costs.

Overhead as a Percentage of Direct Costs		Overhead and Profit Allowance — Add to Items That Do Not Include Subcontractor's O&P — Average	Allowance to Add to Items That Do Include Subcontractor's O&P		Typical by Size of Project	
Minimum	5%	25%	Minimum	5%	under $100,000	30%
Average	12%		Average	10%	$500,000	25%
Maximum	22%		Maximum	15%	$2,000,000	20%
					over $10,000,000	15%

Insurance Rates

Type	Minimum	Maximum
Builder's Risk	.22%	.59%
All-risk Type	.25%	.62%
Contractor's Equipment Floater	.50%	1.50%
Public Liability, Average	—	1.55%

This table represents approximate values relative to total project cost for the most common types of basic insurance coverages.

Builder's Risk Insurance Rates

Builder's Risk Insurance is insurance on a building during construction. Premiums are paid by the owner or the contractor. Blasting, collapse and underground insurance would raise total insurance costs above those listed. Floater policy for materials delivered to the job runs $.75 to $1.25 per $100 value. Contractor equipment insurance runs $.50 to $1.50 per $100 value. Insurance for miscellaneous tools to $1,000 value runs from $5.50 to $7.50 per $100 value.

Tabulated below are New England Builder's Risk insurance rates in dollars per $100 value for $1,000 deductible. For $25,000 deductible, rates can be reduced 13% to 34%. On contracts over $1,000,000, rates may be lower than those tabulated. Policies are written annually for the total completed value in place. For "all risk" insurance (excluding flood, earthquake and certain other perils) add $.025 to total rates below.

Coverage	Frame Construction (Class 1)				Brick Construction (Class 4)				Fire Resistive (Class 6)			
	Range			Average	Range			Average	Range			Average
Fire Insurance	$.300	to	$.420	$.394	$.132	to	$.189	$.174	$.052	to	$.080	$.070
Extended Coverage	.115	to	.150	.144	.080	to	.105	.101	.081	to	.105	.100
Vandalism	.012	to	.016	.015	.008	to	.011	.011	.008	to	.011	.010
Total Annual Rate	$.427	to	$.586	$.553	$.220	to	.305	$.286	$.141	to	$.196	$.180

Permit Rates

Project Permits	Minimum	Maximum
Rule of thumb, most cities	.50%	2%

Permit costs vary greatly depending on many factors, such as type of project, proposed occupancy, location, local codes, need for variances or change of zoning, etc. This table provides a "rule of thumb" to use when local conditions cannot be determined.

Small Tools Allowance

Small Tools Allowance	Minimum	Maximum
As % of contractor's work	.50%	2%

A variety of small tools must be purchased in the course of almost every project, whether to complete small tasks or to replace tools that have "mysteriously disappeared." This table provides a "rule of thumb" that can be used to assign a value to this often overlooked item.

Project Desirability

In the estimating and bidding of construction projects, one of the first questions that you must ask yourself is "Do I really want to bid this job?" The question is simple enough; it's the answer that is sometimes the problem. Judging whether you should bid on a project or pass on it depends on its desirability as viewed by each bidder. Some projects are clear cut in this regard. Those that fall somewhere in the middle are hard to call.

Bid dates (deadlines), dollar size, site location, type of construction, and other pertinent conditions affect the immediate choice. They will continue to influence the deeper analysis that must follow.

Analyzing the project serves three purposes: (1) to reach a decision on whether or not to bid, (2) to determine the degree of effort and competitiveness to apply if you decide to bid, and (3) to establish the amount of the final markup for profit.

In reality, a bidder performs this analysis, referred to hereafter as a project's rating, or desirability rating, somewhat intuitively and with little conscious deliberation. A few typical conditions are:

1. Size (cost of the project in dollars)
2. Location of project (distance of the project from the home office)
3. Sponsor relation
4. Type of construction
5. Probable competitiveness
6. Labor market
7. Subcontractor market
8. Quality of drawings and specifications
9. Quality of supervision
10. Special risks
11. Completion time and penalty
12. Estimating and bidding time
13. Need for work
14. Other special advantages or disadvantages

If all these conditions were positive, the project would be highly desirable, the bidding competitive, and the markup minimal. If most of these conditions were negative, the project would be less desirable, the bidding conservative, and the markup maximum.

The markup is a variable that helps to compensate for the strengths or weaknesses of a project, as summarized in its rating. For instance, the example below shows two projects of equal size with different ratings. The increased markup tends to make Project A more desirable. And yet, if the theory is correct, it would not seriously decrease competitiveness, since other bidders should also make similar compensations for the obvious differences.

	Basic Cost	**Rating**	**Markup**	**Add'l. Cost**
Project A	$42,000	78	*10.2%	$4,284
Project B	$42,000	88	*9.0%	$3,780

*Assumed Theoretical Percentages

This example graphically portrays the principle.

If both projects were equal in ratings, the markups and bids would be equal. The rating works both ways, to raise or to lower a bid relative to some arbitrary average. It is thus a realistic aspect of good bidding practice.

Let us now examine each of the previously named conditions. We cannot expect them to be of equal weight, so there should be different rules for each. Let us reserve all numbers on the 0-100 scale below 30 and above 90 for very exceptional conditions, in order to give them additional impact in the final averaging.

1. **Size of project.** The ideal size is that which will fill up the company's unused bonding capacity. If, for instance, a company has a nominal bonding capacity of $200,000 and $120,000 of uncompleted work on hand, then an $80,000 project would be an ideal size for bidding.
   Let us rate projects for size between the two following extremes:

Ideal size	90
Smaller than ideal	50

2. **Location of project.** For the distance of the project from the home office, suggested values are:

Less than 5 miles	90
More than 30 miles	60

3. **Owner relations.** Experienced bidders know that the efficiency and cooperativeness of the owner's agents contribute to the desirability rating of a project and can influence the decision of the bidder on whether to bid or not. Suggested values are:

Known good relations	90
Unknown	70
Known poor relations	50

4. **Type of construction.** As a rule, a contractor is not equally experienced and capable in all types of construction. The type of construction in which the company excels is the most desirable to a bidder and determines the rating between the following extremes:

Experienced in this type	90
No experience in this type	60

5. **Probable.** The bidder looks with less favor on those projects that attract a large number of bidders and excessive competition. This aspect of a project can carry a lot of weight with values suggested as follows:

Light competition	90
Strong competition	60

6. **Labor market conditions.** Projects located in areas where there are an insufficient number of skilled workers are of low desirability to bidders. Suggested values are:

Good quality and plentiful	90
Poor quality and few	60

7. **Subcontract market conditions.** The quantity and quality of available local subcontractors parallels that of workers, and the suggested values are:

Good quality and plentiful	90
Poor quality and few	60

8. **Quality of drawings and specifications.** Quality is not a heavy factor, but it does affect the desirability of a project. Suggested values are:

Good quality	90
Poor quality	70

9. **Quality of supervision.** When the identity of the superintendent is known and of proven excellence, the project is more desirable to the bidder. An unknown superintendent, or one of mediocre ability, is a negative influence in the rating. The following values are suggested:

Known top quality	90
Unknown	80
Known mediocre quality	70

10. **Special risks.** Conditions such as solid rock, subsurface water, high altitude, extreme heat or cold, possibility of collapse, cave-in, etc., all affect the desirability rating. Depending on the degree of risk, values may be scaled between:

No known risk	90
Extreme risk	30

11. **Completion time and penalty.** When the time provided in the bid documents to complete the construction is extremely short and the penalty for delay is very high, the project loses desirability to the bidder. Values are:

Time sufficient and penalty low	90

Time sufficient and penalty high	60

12. **Estimating and bidding time.** If the bidder is not given sufficient time to put the bid together, a low level of desirability is assigned to the project. This condition might also exist because of the coincidence of two projects bidding on or near the same date. Values are:

Sufficient time to bid	90
Insufficient time to bid	50

13. **Need for work.** This condition could carry more weight than any of the others. If extreme, the bidder's need for work might reverse an otherwise unfavorable rating. Suggested values are:

Extreme need for work	100
Normal need	85
Very little need	50

14. **Other special advantages or disadvantages.** A bidder might have special incentives, positive or negative, such as the exclusive possession of a stockpile of materials. Or the bidder might have knowledge of a competitor's possession of such an advantage. The weight for this condition can vary considerably, but if there are no special advantages or disadvantages, an average may be used, as follows:

Strong advantages	100
Average	80
Strong disadvantages	25

In a similar way, but with fewer facts, the bidder roughly analyzes the competition's probable degree of interest in the project. The result may cause a further final adjustment in the bidder's own evaluation.

The following figure is an example of a hypothetical project rated for desirability.

1. Size of project in dollars	90
2. Location of project (distance from home office)	90
3. Owner, architect, inspector (sponsor) relations	80
4. Type of construction (experienced or not)	90
5. Competition	80
6. Labor market	75
7. Subcontractor market	80
8. Quality of drawings and specifications	85
9. Quality of supervision	90
10. Special risks	25
11. Completion time and liquidated damages	85
12. Time to estimate and bid	80
13. Need of a job (contract)	100
14. Other	85
	1,135

$$\frac{1{,}135}{14} = 81.07 \text{ average}$$

Two extremes are shown: special risk and need for work. These two exceptional conditions tend to balance one another and make the project of average desirability, according to the following scale:

Very desirable	85 to 90
Average	80 to 85
Undesirable	Below 80

Remember, a project may be made more desirable by an increase in the markup and by contingency allowances.

Basics of Project Management

Anyone who has spent an hour on a job site, from the seasoned old pro to the greenest young laborer, knows that any construction project is a monument to Murphy's Law, ("Whatever **can** go wrong, **will** go wrong."), and its corollary, ("It will go wrong at the worst possible time."). A construction project can be viewed as an ongoing exercise in creative problem-solving on the part of everyone involved, with the buck ultimately stopping on the desk of the project manager. It can be truly said that success in the construction business is in direct correlation to your ability to master the skills of project management.

Construction project management can be defined as the planning, staffing, directing, and controlling of a company's resources in order to achieve the goals set for a construction project. What follows is a brief outline of the basic tasks involved in successful project management.

1. **Estimate** as accurately as possible the cost of the project. This should include a detailed breakdown of all phases ("systems") of the project in terms of time ("labor-hours") and materials ("components"). No management skill has more of a direct impact on your profitability than your ability to produce an accurate estimate.
2. **Establish specific goals** in terms of how much time and money you wish to spend. These do not necessarily have to correspond exactly to your estimate. For example, your estimate might call for framing to be completed in six weeks, but you set your goal for five weeks. Your overall goal is always not to exceed your estimate. Coming in under your estimate is money in your pocket.
3. **Plan ahead** on how you are going to achieve your goals. This may involve anything from deciding how stock is to be arranged on the job site to coordinating the purchase of materials to take advantage of price breaks offered by different suppliers. Planning takes time and effort, but it is as much a part of a successful builder's skills as the ability to drive a nail. Anticipating problems and solving them through careful and imaginative planning is the most effective way of ensuring that the physical job of building will go smoothly and result in good work at minimal cost. You cannot completely escape the clutches of Murphy's Law, but good planning practices will generally bail you out.
4. **Communicate** your goals and plans to those who will execute or be affected by them. Be certain your crew understands what you expect them to produce and how and when you expect them to produce it. Give your directions clearly and encourage them to ask questions whenever they are unsure of what their job is or how it is to be done. Keep your client up to date on the progress of the job and solicit his or her comments and suggestions, particularly regarding design and finish treatment. This can often be an annoying task and a trial of your patience, but keeping your client involved in the process and satisfied with the results is a very important factor in your formula for success. The client, after all, writes your paycheck

and represents the best, and cheapest, advertisements for your business.

5. **Establish** a chain of command. Be sure everyone involved in the project understands who is in charge of, or answerable to, whom. From architects and designers, through foremen and subcontractors, down to the laborer who sweeps up the sawdust — all must be aware of how decisions are made and who is called upon to make them. This is often a self-evident matter within the various crews, but it can get extremely complicated when a problem or design change comes up which may involve one architect, two engineers, and three tradesmen, all of whom seem to be at cross-purposes. You must then bring to bear all your skills as a communicator, mediator, and problem-solver to come to a decision that satisfies everyone.
6. **Coordinate** the use of resources to meet your goals. Simply put, this means putting your plans into action. You have spent time deciding on the scheduling of operations, the building sequence, and the most efficient and economical use of materials, labor, tools, equipment, and outside services. Now it is up to you to see that everything gets done as planned. The success of your project very much depends on taking those ideas you put on paper and putting them into practice on the job. After all, there is no sense in writing a stock list if you don't read it when you get to the lumberyard.
7. **Monitor and document** all activities and events that affect, directly or indirectly, the work that you and/or your company control. Maintain a project journal in a notebook. Set aside a little time each day to record a brief account of the job's progress: goals and deadlines met or missed; problems confronted: their causes and solutions; personnel difficulties; business conversations; promises or commitments received or given; in short, anything of significance to the orderly completion of the project. You will thus have, for immediate reference, an up-to-date assessment of the state of the project, as well as the information necessary to make intelligent decisions and adjustments to keep things running smoothly. Journals of past projects can shed light on current problems and will often contain insights, answers, solutions, and information you forgot you knew. As part of the same journal, or perhaps in a separate notebook, maintain an updated list of "extras," that is, labor and materials not included in the original estimate. Write down a description of each extra, noting costs, and have it initialed by the client, designer, subcontractor, or any other person in charge of carrying out the work. This ensures that all interested parties are aware and informed of all extras and of their specific obligations with regard to them. Confusion, misunderstanding, and hard feelings are thus forestalled, and a professional tone is maintained on the job that will be to everyone's benefit and help to produce quality results.

Overall project management can be more fully illustrated in the following figure–the Project Control Cycle. Notice that the feedback loop is part of the overall project management responsibility.

The steps involved in the Project Control Cycle are:

STEP 1 — Set Initial Goals. This involves creating an estimate of cost and time needed for the project. In many cases these are already established for you by way of a bid and/or a contract.

STEP 2 — Establish the Job Plan. Build a schedule and create a budget of how much you want to spend — not how much you can spend.

STEP 3 — Monitor the Progress. Collect data on how much money has been spent, how much time has been used, and the overall state of the job.

STEP 4 — Process the Information. Compare the data collected in Step 3 against the job plan.

STEP 5 — Compare and Analyze. Look for deviations from the job plan. Determine what the causes are.

STEP 6 — Take Corrective Action. Decide what needs to be done and do it, or have it done. (NOTE: If you do not take corrective action when it is warranted then the whole concept of project management falls apart. If you are not prepared to take corrective action, no management system will work.)

STEP 7 — Collect Historical Data. The information gathered on any project serves many purposes. It will: 1) be extremely useful when estimating your next job; 2) help prevent you from falling into a bad situation again; 3) assist you in getting out of a similar situation if it should occur again; and 4) document all events clearly as they occur, in case legal action is taken by or against you.

We cannot guarantee that all of your projects will be successful by using project management procedures, but, if used properly, such procedures will go a long way toward making your projects run smoother and your business more profitable.

Scheduling Concerns

There are many factors that must be taken into consideration when preparing a construction schedule. They include the time allowed (in a contract, for instance), job location, the climate of the region, the time of the year, deliveries and availabilities of materials, the crew make-up, size and complexity of the job, administrative requirements, sample and document submittals for approval, and the coordination with other trades.

Scheduling is determining the sequence of events that must occur to complete your project. To create a schedule, you must examine each task of your project and ask these three questions:

What task must be completed before this activity can start?
What task can be worked at the same time as this task?
What task cannot start until this activity is complete?

When the order of work is determined by the above process, you have a basic schedule completed. You now need to figure out how long each task will take to complete. Work with your superintendents, foremen and/or the workers. They have the working knowledge of how long it takes to complete a task. If you are doing work that you and/or your crews are not familiar with, then consult the Means labor-hour columns to give you a good idea of how long it will take.

If the time allowed for you to complete the work is already spelled out for you, you at least have a framework of time to work within and can form your crews to fit accordingly.

Allow for weather delays in regions and/or times of year they are warranted. In New England, for outside work as a general rule, add two days per week for winter work, add one day per week for the spring and the fall seasons, and one day for every two weeks of work in the summertime.

You also need to take into consideration the availability of workers in the trades; whether they live near the project; whether they'll stay on or jump to another job, or whether they have experience in this type of work.

A last item concerning scheduling: Do not let the schedule absolutely dictate how the job is to be run. If a long lead item suddenly shows up on a site a week early, don't determine that it has to sit in a storage trailer until the schedule says it's okay to install it. These are the pleasant type of surprises we rarely get! Rework your schedule to determine if you can save any time by installing it **now**. Again, run through various "what if...?" types of situations. In other words, be firm — maintain the schedule at all costs; but be flexible enough and open to the inevitable changes that will occur on your job site. If your information is current and logic is sound, you should be able to take advantage of any situation that occurs and turn it into a gain for yourself and your project.

Section Four
Construction Forms

Forms are an important management tool. Consistent use of a specific set of forms helps to organize and standardize the collection and reporting of information. As a result, those who are involved in data gathering and analysis become so familiar with the use of the forms that they can more easily focus on the information without being distracted by the form itself. Estimate forms are designed to record and summarize the quantity calculations and systems pricing on a single sheet. There are columns to list descriptive information as well as material and installation costs on aper unit basis. Additional columns are provided to calculate project specific costs after determination of total square footage or total number of units appropriate to the system being estimated. Totals are then determined by adding the extended figures across and down the form.

The forms in this section are ready for direct use. After the appropriate form has been selected, the form can be reproduced on an office copier as needed. Purchase of this book grants the owner permission to reproduce unlimited quantities of these forms for his/her own use but not for resale.

Sample Form

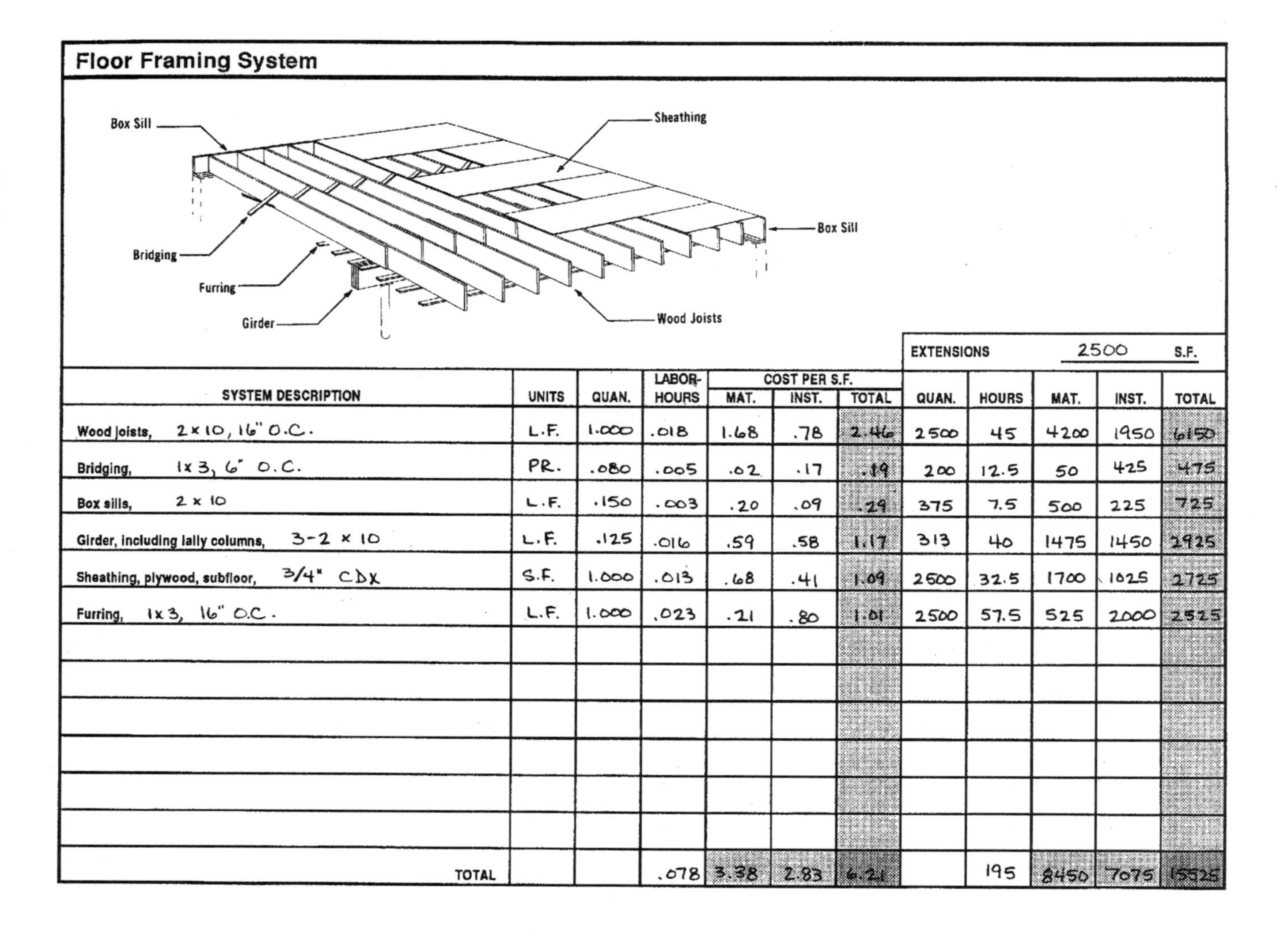

Floor Framing System

EXTENSIONS 2500 S.F.

SYSTEM DESCRIPTION	UNITS	QUAN.	LABOR-HOURS	COST PER S.F. MAT.	INST.	TOTAL	QUAN.	HOURS	MAT.	INST.	TOTAL
Wood joists, 2×10, 16" O.C.	L.F.	1.000	.018	1.68	.78	2.46	2500	45	4200	1950	6150
Bridging, 1x3, 6" O.C.	PR.	.080	.005	.02	.17	.19	200	12.5	50	425	475
Box sills, 2 x 10	L.F.	.150	.003	.20	.09	.29	375	7.5	500	225	725
Girder, including lally columns, 3-2 x 10	L.F.	.125	.016	.59	.58	1.17	313	40	1475	1450	2925
Sheathing, plywood, subfloor, 3/4" CDX	S.F.	1.000	.013	.68	.41	1.09	2500	32.5	1700	1025	2725
Furring, 1x3, 16" O.C.	L.F.	1.000	.023	.21	.80	1.01	2500	57.5	525	2000	2525
TOTAL			.078	3.38	2.83	6.21		195	8450	7075	15525

Floor Framing System

							EXTENSIONS				____ S.F.
			LABOR-	COST PER S.F.							
SYSTEM DESCRIPTION	UNITS	QUAN.	HOURS	MAT.	INST.	TOTAL	QUAN.	HOURS	MAT.	INST.	TOTAL
Wood Joists,											
Bridging,											
Box sills,											
Girder, including lally columns,											
Sheathing, plywood, subfloor,											
Furring,											
TOTAL											

Floor Framing System

CWJ Rim Joist
Plywood Sheathing
Temporary Strut Lines 1" x 4", 8'-0" O.C.
Web Stiffener
Laminated Veneer Lumber Beam, (LVL)
Composite Wood Joists (CWJ)

SYSTEM DESCRIPTION	UNITS	QUAN.	LABOR-HOURS	COST PER S.F.			EXTENSIONS ______ S.F.				
				MAT.	INST.	TOTAL	QUAN.	HOURS	MAT.	INST.	TOTAL
Composite Wood Joists											
Temp. strut line											
CWJ rim joist											
Girder, including lally columns,											
Sheathing, plywood, subfloor,											
TOTAL											

Floor Framing System

Cont. 2" x 4" Ribbon

Plywood Sheathing

Girder

Wood Floor Trusses

SYSTEM DESCRIPTION	UNITS	QUAN.	LABOR-HOURS	COST PER S.F.			EXTENSIONS ____ S.F.				
				MAT.	INST.	TOTAL	QUAN.	HOURS	MAT.	INST.	TOTAL
Open Web Wood joists											
Continuous ribbing											
Girder, including lally columns,											
Sheathing, plywood, subfloor,											
Furring,											
TOTAL											

Exterior Wall Framing System

Sheathing
Top Plates
Studs
Bottom Plate
Corner Bracing

SYSTEM DESCRIPTION	UNITS	QUAN.	LABOR-HOURS	COST PER S.F.			EXTENSIONS S.F.				
				MAT.	INST.	TOTAL	QUAN.	HOURS	MAT.	INST.	TOTAL
Wood studs,											
Plates, double top, single bottom,											
Corner bracing,											
Sheathing,											
TOTAL											

Gable End Roof Framing System

SYSTEM DESCRIPTION	UNITS	QUAN.	LABOR-HOURS	COST PER S.F.			EXTENSIONS ______ S.F.				
				MAT.	INST.	TOTAL	QUAN.	HOURS	MAT.	INST.	TOTAL
Wood rafters,											
Ceiling joists,											
Ridge board,											
Fascia board,											
Rafter tie,											
Soffit nailer,											
Sheathing,											
Furring strips,											
TOTAL											

Truss Roof Framing System

Sheathing
Trusses
Fascia Board
Furring

SYSTEM DESCRIPTION	UNITS	QUAN.	LABOR-HOURS	COST PER S.F.			EXTENSIONS ______ S.F.				
				MAT.	INST.	TOTAL	QUAN.	HOURS	MAT.	INST.	TOTAL
Truss, loading,											
Fascia board,											
Sheathing,											
Furring,											
TOTAL											

Hip Roof Framing System

SYSTEM DESCRIPTION	UNITS	QUAN.	LABOR-HOURS	COST PER S.F.			EXTENSIONS _______ S.F.				
				MAT.	INST.	TOTAL	QUAN.	HOURS	MAT.	INST.	TOTAL
Hip rafters,											
Jack rafters,											
Ceiling joists,											
Fascia board,											
Soffit nailer,											
Sheathing,											
Furring strips,											
TOTAL											

Gambrel Roof Framing System

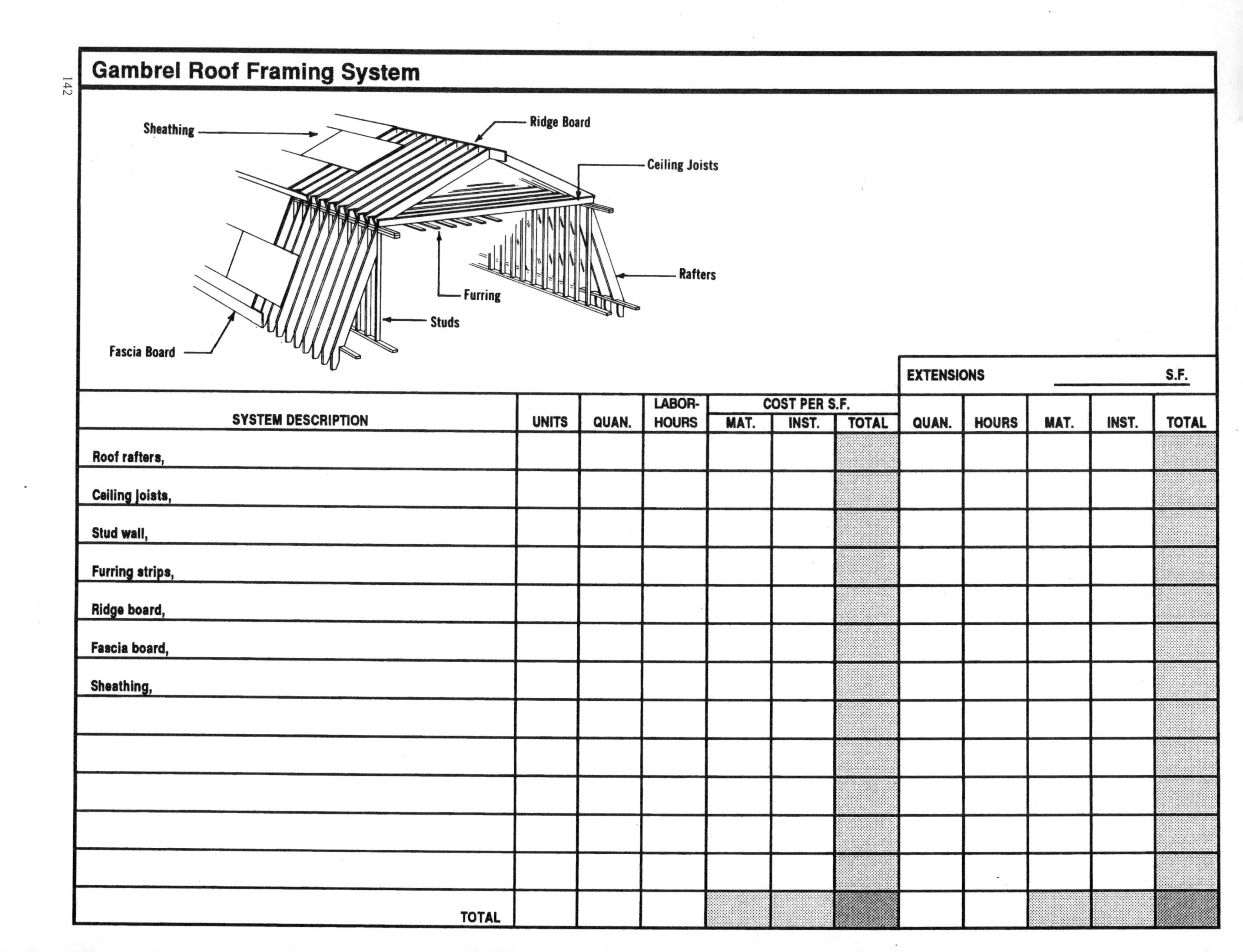

SYSTEM DESCRIPTION	UNITS	QUAN.	LABOR-HOURS	COST PER S.F.			EXTENSIONS ____ S.F.				
				MAT.	INST.	TOTAL	QUAN.	HOURS	MAT.	INST.	TOTAL
Roof rafters,											
Ceiling joists,											
Stud wall,											
Furring strips,											
Ridge board,											
Fascia board,											
Sheathing,											
TOTAL											

Mansard Roof Framing System

SYSTEM DESCRIPTION	UNITS	QUAN.	LABOR-HOURS	COST PER S.F.			EXTENSIONS ____ S.F.				
				MAT.	INST.	TOTAL	QUAN.	HOURS	MAT.	INST.	TOTAL
Roof rafters,											
Rafter plates,											
Ceiling joists,											
Hip rafter,											
Jack rafter,											
Ridge board,											
Sheathing,											
Furring strips,											
TOTAL											

Shed/Flat Roof Framing System

EXTENSIONS ________ S.F.

SYSTEM DESCRIPTION	UNITS	QUAN.	LABOR-HOURS	COST PER S.F.			EXTENSIONS				
				MAT.	INST.	TOTAL	QUAN.	HOURS	MAT.	INST.	TOTAL
Roof rafters,											
Fascia board,											
Bridging,											
Sheathing,											
TOTAL											

Gable Dormer Framing System

SYSTEM DESCRIPTION	UNITS	QUAN.	LABOR-HOURS	COST PER S.F.			EXTENSIONS S.F.				
				MAT.	INST.	TOTAL	QUAN.	HOURS	MAT.	INST.	TOTAL
Dormer rafter,											
Ridge board,											
Trimmer rafters,											
Wall, studs and plates,											
Fascia board,											
Valley rafter,											
Cripple rafter,											
Headers,											
Ceiling joists,											
Sheathing,											
TOTAL											

Shed Dormer Framing System

Sheathing
Ceiling Joists
Fascia Board
Rafters
Studs & Plates
Trimmer Rafters

SYSTEM DESCRIPTION	UNITS	QUAN.	LABOR-HOURS	COST PER S.F.			EXTENSIONS ____ S.F.				
				MAT.	INST.	TOTAL	QUAN.	HOURS	MAT.	INST.	TOTAL
Dormer rafter,											
Trimmer rafter,											
Wall, studs and plates,											
Fascia board,											
Ceiling joists,											
Sheathing,											
TOTAL											

Partition Framing System

Bracing
Studs
Top Plates
Bottom Plate

												EXTENSIONS _______ S.F.
SYSTEM DESCRIPTION	UNITS	QUAN.	LABOR-HOURS	COST PER S.F.								
				MAT.	INST.	TOTAL	QUAN.	HOURS	MAT.	INST.	TOTAL	
Wood studs,												
Plates, double top, single bottom,												
Cross bracing,												
TOTAL												

Wood Siding System

Trim

Building Paper

Beveled Cedar Siding

SYSTEM DESCRIPTION	UNITS	QUAN.	LABOR-HOURS	COST PER S.F.			EXTENSIONS ______ S.F.				
				MAT.	INST.	TOTAL	QUAN.	HOURS	MAT.	INST.	TOTAL
Siding,											
Building paper,											
Trim,											
Finish,											
TOTAL											

Shingle Siding System

Trim

Building Paper

White Cedar Shingles

SYSTEM DESCRIPTION	UNITS	QUAN.	LABOR-HOURS	COST PER S.F.			EXTENSIONS ______ S.F.				
				MAT.	INST.	TOTAL	QUAN.	HOURS	MAT.	INST.	TOTAL
Shingles,											
Building paper,											
Trim,											
Finish,											
TOTAL											

Metal and Plastic Siding System

Alum. Trim
Building Paper
Alum. Horizontal Siding
Backer Insulation Board

SYSTEM DESCRIPTION	UNITS	QUAN.	LABOR-HOURS	COST PER S.F.			EXTENSIONS ______ S.F.				
				MAT.	INST.	TOTAL	QUAN.	HOURS	MAT.	INST.	TOTAL
Siding,											
Backer, insulation board,											
Building paper,											
Trim,											
TOTAL											

Double Hung Window System

Drip Cap
Snap-In-Grille
Caulking
Interior Trim
Window

SYSTEM DESCRIPTION	UNITS	QUAN.	LABOR-HOURS	COST EACH			EXTENSIONS				EACH
				MAT.	INST.	TOTAL	QUAN.	HOURS	MAT.	INST.	TOTAL
Window,											
Trim,											
Finish,											
Caulking,											
Grille,											
Drip cap,											
TOTAL											

Casement Window System

Drip Cap
Snap-In-Grille
Interior Trim
Caulking
Window

SYSTEM DESCRIPTION	UNITS	QUAN.	LABOR-HOURS	COST EACH			EXTENSIONS ______ EACH				
				MAT.	INST.	TOTAL	QUAN.	HOURS	MAT.	INST.	TOTAL
Window,											
Trim,											
Finish,											
Caulking,											
Grille,											
Drip cap,											
TOTAL											

Awning Window System

Drip Cap
Interior Trim
Snap-In-Grille
Caulking
Window

			LABOR-	COST EACH			EXTENSIONS				EACH
SYSTEM DESCRIPTION	UNITS	QUAN.	HOURS	MAT.	INST.	TOTAL	QUAN.	HOURS	MAT.	INST.	TOTAL
Window,											
Trim,											
Finish,											
Caulking,											
Grille,											
Drip cap,											
TOTAL											

Sliding Window System

SYSTEM DESCRIPTION	UNITS	QUAN.	LABOR-HOURS	COST EACH			EXTENSIONS			EACH	
				MAT.	INST.	TOTAL	QUAN.	HOURS	MAT.	INST.	TOTAL
Window,											
Trim,											
Finish,											
Caulking,											
Grille,											
Drip cap,											
TOTAL											

Bow/Bay Window System

Snap-In-Grille
Drip Cap
Caulking
Window

SYSTEM DESCRIPTION	UNITS	QUAN.	LABOR-HOURS	COST EACH			EXTENSIONS			EACH	
				MAT.	INST.	TOTAL	QUAN.	HOURS	MAT.	INST.	TOTAL
Window,											
Trim,											
Finish,											
Caulking,											
Grille,											
Drip cap,											
TOTAL											

Fixed Window System

Drip Cap
Interior Trim
Caulking
Snap-In-Grille
Window

SYSTEM DESCRIPTION	UNITS	QUAN.	LABOR-HOURS	COST EACH			EXTENSIONS ______ EACH				
				MAT.	INST.	TOTAL	QUAN.	HOURS	MAT.	INST.	TOTAL
Window,											
Trim,											
Finish,											
Caulking,											
Grille,											
Drip cap,											
TOTAL											

Entrance Door System

Drip Cap

Door

Frame & Exterior Casing

Interior Casing

Sill

SYSTEM DESCRIPTION	UNITS	QUAN.	LABOR-HOURS	COST EACH			EXTENSIONS EACH				
				MAT.	INST.	TOTAL	QUAN.	HOURS	MAT.	INST.	TOTAL
Door,											
Frame and exterior casing,											
Interior casing,											
Sill,											
Hinges,											
Lockset,											
Weatherstripping,											
Finish,											
TOTAL											

Sliding Door System

Drip Cap
Frame & Exterior Casing
Door
Sill
Interior Casing

SYSTEM DESCRIPTION	UNITS	QUAN.	LABOR-HOURS	COST EACH MAT.	COST EACH INST.	COST EACH TOTAL	EXTENSIONS QUAN.	EXTENSIONS HOURS	EXTENSIONS MAT.	EXTENSIONS INST.	EXTENSIONS TOTAL
Door,											
Frame and exterior casing,											
Interior casing,											
Sill,											
Finish,											
Drip cap,											
TOTAL											

EXTENSIONS ______ EACH

Residential Garage Door System

Exterior Trim
Door
Jamb
Drip Cap
Weatherstripping

SYSTEM DESCRIPTION	UNITS	QUAN.	LABOR-HOURS	COST EACH			EXTENSIONS			EACH	
				MAT.	INST.	TOTAL	QUAN.	HOURS	MAT.	INST.	TOTAL
Door,											
Jamb and header blocking,											
Exterior trim,											
Finish,											
Weatherstripping,											
Drip cap,											
TOTAL											

Aluminum Window System

SYSTEM DESCRIPTION	UNITS	QUAN.	LABOR-HOURS	COST EACH			EXTENSIONS			EACH	
				MAT.	INST.	TOTAL	QUAN.	HOURS	MAT.	INST.	TOTAL
Window,											
Blocking,											
Drywall,											
Corner bead,											
Finish drywall,											
Sill,											
TOTAL											

Gable End Roofing System

Ridge Shingles
Building Paper
Shingles
Rake Board
Drip Edge
Gutter
Soffit & Fascia
Downspouts

SYSTEM DESCRIPTION	UNITS	QUAN.	LABOR-HOURS	COST PER S.F.			EXTENSIONS _____ S.F.				
				MAT.	INST.	TOTAL	QUAN.	HOURS	MAT.	INST.	TOTAL
Shingles,											
Drip edge,											
Building paper,											
Ridge shingles,											
Soffit and fascia,											
Rake trim,											
Gutter,											
Downspouts,											
TOTAL											

Hip Roof Roofing System

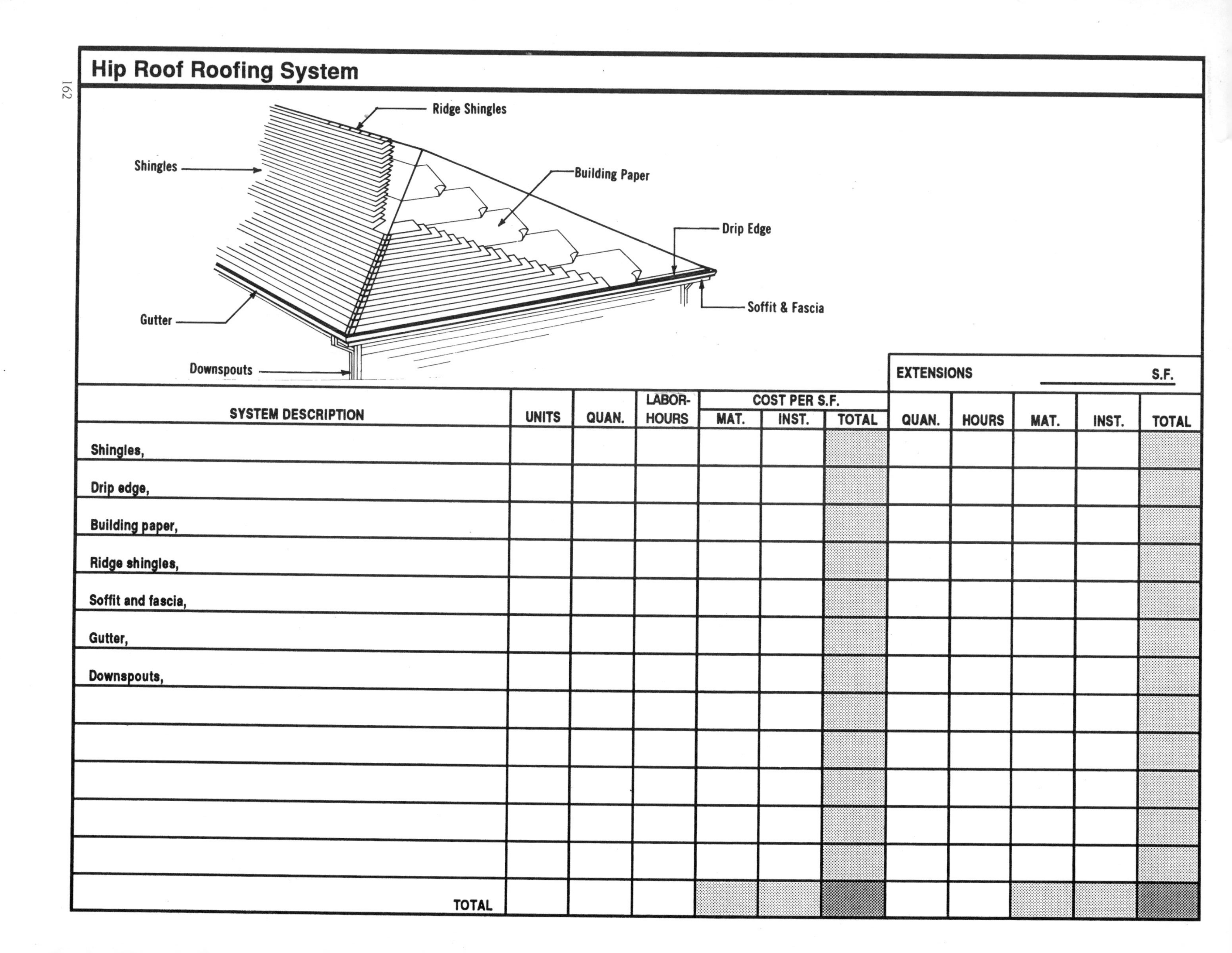

SYSTEM DESCRIPTION	UNITS	QUAN.	LABOR-HOURS	COST PER S.F.			EXTENSIONS ____ S.F.				
				MAT.	INST.	TOTAL	QUAN.	HOURS	MAT.	INST.	TOTAL
Shingles,											
Drip edge,											
Building paper,											
Ridge shingles,											
Soffit and fascia,											
Gutter,											
Downspouts,											
TOTAL											

Gambrel Roofing System

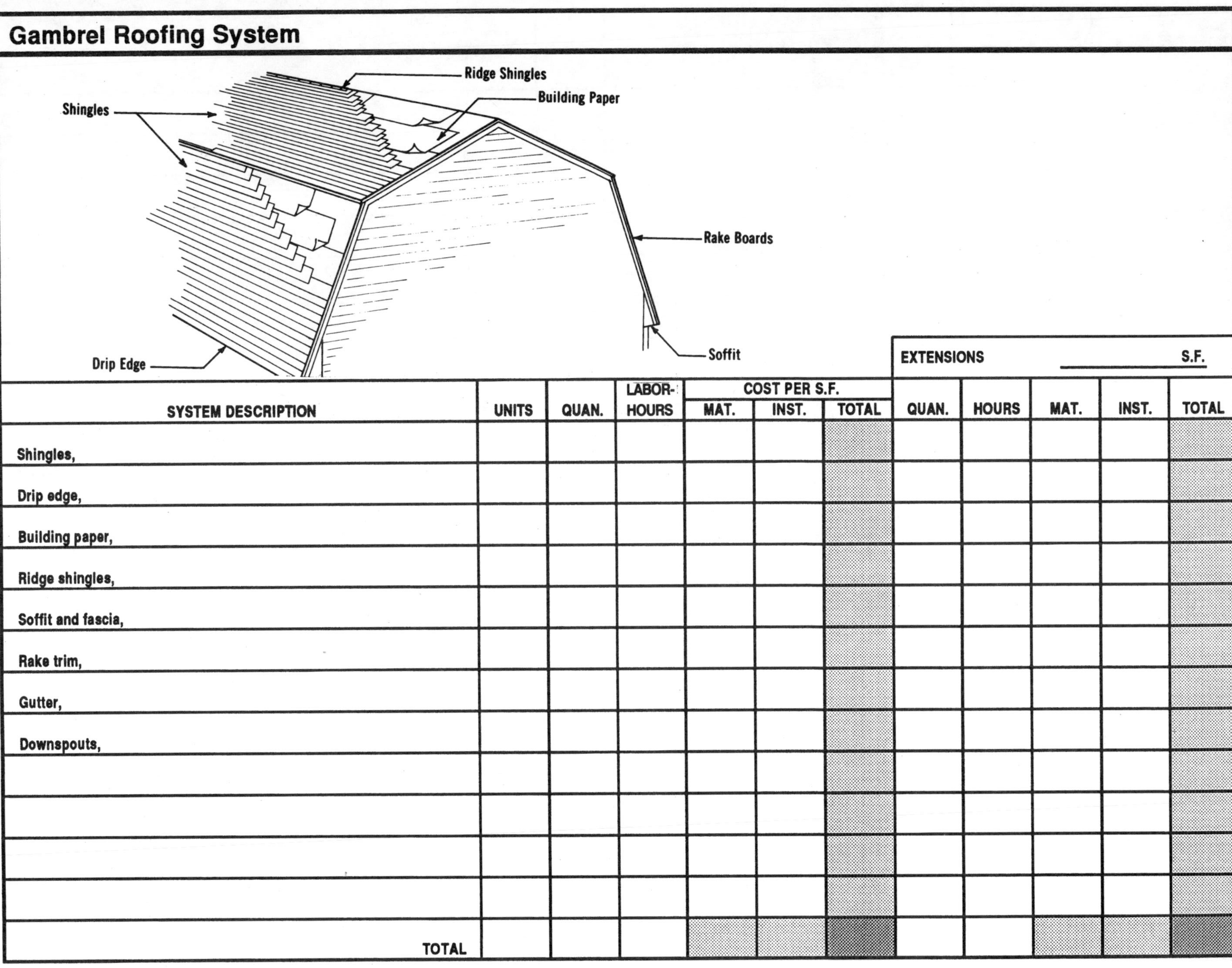

SYSTEM DESCRIPTION	UNITS	QUAN.	LABOR-HOURS	COST PER S.F.			EXTENSIONS ______ S.F.				
				MAT.	INST.	TOTAL	QUAN.	HOURS	MAT.	INST.	TOTAL
Shingles,											
Drip edge,											
Building paper,											
Ridge shingles,											
Soffit and fascia,											
Rake trim,											
Gutter,											
Downspouts,											
TOTAL											

Mansard Roofing System

SYSTEM DESCRIPTION	UNITS	QUAN.	LABOR-HOURS	COST PER S.F.			EXTENSIONS S.F.				
				MAT.	INST.	TOTAL	QUAN.	HOURS	MAT.	INST.	TOTAL
Shingles,											
Drip edge,											
Building paper,											
Ridge shingles,											
Soffit and fascia,											
Gutter,											
Downspouts,											
TOTAL											

Shed Roofing System

Building Paper
Drip Edge
Shingles
Soffit & Fascia
Gutter
Rake Boards
Downspouts

SYSTEM DESCRIPTION	UNITS	QUAN.	LABOR-HOURS	COST PER S.F.			EXTENSIONS S.F.				
				MAT.	INST.	TOTAL	QUAN.	HOURS	MAT.	INST.	TOTAL
Shingles,											
Drip edge,											
Building paper,											
Soffit and fascia,											
Rake trim,											
Gutter,											
Downspouts,											
TOTAL											

Gable Dormer Roofing System

Ridge Shingles
Building Paper
Shingles
Rake Boards
Drip Edge
Soffit & Fascia
Flashing

SYSTEM DESCRIPTION	UNITS	QUAN.	LABOR-HOURS	COST PER S.F.			EXTENSIONS _____ S.F.				
				MAT.	INST.	TOTAL	QUAN.	HOURS	MAT.	INST.	TOTAL
Shingles,											
Drip edge,											
Building paper,											
Ridge shingles,											
Soffit and fascia,											
Flashing,											
TOTAL											

Shed Dormer Roofing System

Building Paper
Shingles
Drip Edge
Rake Boards
Soffit & Fascia
Flashing

SYSTEM DESCRIPTION	UNITS	QUAN.	LABOR-HOURS	COST PER S.F.			EXTENSIONS ______ S.F.				
				MAT.	INST.	TOTAL	QUAN.	HOURS	MAT.	INST.	TOTAL
Shingles,											
Drip edge,											
Building paper,											
Soffit and fascia,											
Flashing,											
TOTAL											

Skylight/Skywindow System

Skylight
Flashing
Curb
Trimmer Rafter
Interior Trim
Headers

SYSTEM DESCRIPTION	UNITS	QUAN.	LABOR-HOURS	COST EACH MAT.	COST EACH INST.	COST EACH TOTAL	EXTENSIONS (EACH) QUAN.	HOURS	MAT.	INST.	TOTAL
Skylight or skywindow,											
Trimmer rafters,											
Headers,											
Curb,											
Flashing,											
Interior trim,											
TOTAL											

Drywall and Thincoat Wall System

SYSTEM DESCRIPTION	UNITS	QUAN.	LABOR-HOURS	COST PER S.F.			EXTENSIONS ______ S.F.				
				MAT.	INST.	TOTAL	QUAN.	HOURS	MAT.	INST.	TOTAL
Drywall,											
Finish,											
Corners,											
Painting,											
Trim,											
TOTAL											

Drywall and Thincoat Ceiling System

Finish
Drywall
Paint
Corners

SYSTEM DESCRIPTION	UNITS	QUAN.	LABOR-HOURS	COST PER S.F.			EXTENSIONS ______ S.F.				
				MAT.	INST.	TOTAL	QUAN.	HOURS	MAT.	INST.	TOTAL
Drywall,											
Finish,											
Corners,											
Painting,											
TOTAL											

Plaster and Stucco Wall System

Plaster
Paint
Trim
Lath

SYSTEM DESCRIPTION	UNITS	QUAN.	LABOR-HOURS	COST PER S.F.			EXTENSIONS ______ S.F.				
				MAT.	INST.	TOTAL	QUAN.	HOURS	MAT.	INST.	TOTAL
Plaster,											
Lath,											
Corners,											
Painting,											
Trim,											
TOTAL											

Plaster and Stucco Ceiling System

Corners
Plaster
Lath
Paint

SYSTEM DESCRIPTION	UNITS	QUAN.	LABOR-HOURS	COST PER S.F.			EXTENSIONS ______ S.F.				
				MAT.	INST.	TOTAL	QUAN.	HOURS	MAT.	INST.	TOTAL
Plaster,											
Lath,											
Corners,											
Painting,											
TOTAL											

Interior Door System

Door
Lockset
Trim
Frame

SYSTEM DESCRIPTION	UNITS	QUAN.	LABOR-HOURS	COST EACH			EXTENSIONS ______ EACH				
				MAT.	INST.	TOTAL	QUAN.	HOURS	MAT.	INST.	TOTAL
Door,											
Frame,											
Trim,											
Hinges,											
Lockset,											
Finish,											
TOTAL											

Closet Door System

Door
Trim
Frame

SYSTEM DESCRIPTION	UNITS	QUAN.	LABOR-HOURS	COST EACH			EXTENSIONS				EACH
				MAT.	INST.	TOTAL	QUAN.	HOURS	MAT.	INST.	TOTAL
Door,											
Frame,											
Trim,											
Finish,											
TOTAL											

Section Five
Reference Aids

This section consists of reference tables and explanations that can be used to develop stock lists and calculate material quantities. The tables contain a wealth of statistical information which you can use to help make your construction projects more economical and profitable. Also included is a list of the abbreviations used in this book.

Number of Board Feet in Various Sized Pieces of Lumber

This table shows how many board feet are in various pieces of lumber for the lengths indicated.

Size (in inches)	Length of Piece							
	8′	10′	12′	14′	16′	18′	20′	22′
2 x 3	4	5	6	7	8	9	10	11
2 x 4	5-1/3	6-2/3	8	9-1/3	10-2/3	12	13-1/3	14-2/3
2 x 6	8	10	12	14	16	18	20	22
2 x 8	10-2/3	13-1/3	16	18-2/3	21-1/3	24	26-2/3	29-1/3
2 x 12	16	20	24	28	32	36	40	44
3 x 4	8	10	12	14	16	18	20	22
3 x 6	12	15	18	21	24	27	30	33
3 x 8	16	20	24	28	32	36	40	44
3 x 10	20	25	30	35	40	45	50	55
3 x 12	24	30	36	42	48	54	60	66
4 x 4	10-2/3	13-1/3	16	18-2/3	21-1/3	24	26-2/3	29-1/3
4 x 6	16	20	24	28	32	36	40	44
4 x 8	21-1/3	26-2/3	32	37-1/3	42-2/3	48	53-1/3	58-2/3
6 x 6	24	30	36	42	48	54	60	66
6 x 8	32	40	48	56	64	72	80	88
6 x 10	40	50	60	70	80	90	100	110
8 x 8	42-2/3	53-1/3	64	74-2/3	85-1/3	96	106-2/3	117-1/3
8 x 10	53-1/3	66-2/3	80	93-1/3	106-2/3	120	133-1/3	146-2/3
8 x 12	64	80	96	112	128	144	160	176
10 x 10	66-2/3	83-1/3	100	116-2/3	133-1/3	150	166-2/3	183-1/3
10 x 12	80	100	120	140	160	180	200	220
12 x 14	112	140	168	196	224	252	280	308

Board Foot Multipliers

This table is used for computing the number of board feet in any length of dimension lumber. Find the size of lumber to be used, and read across for the multiplier. For example, to find out how many board feet of lumber are in an 8′ length of 2 x 4 lumber, look for the multiplier for 2 x 4's (0.667) and multiply it by the length of lumber (8′). The result is 0.667 x 8, or 5.334 board feet.

Nominal Size (in inches)	Multiply Length by	Nominal Size (in inches)	Multiply Length by
2 x 2	0.333	4 x 4	1.333
2 x 3	0.500	4 x 6	2.000
2 x 4	0.667	4 x 8	2.667
2 x 6	1.000	4 x 10	3.333
2 x 8	1.333	4 x 12	4.000
2 x 10	1.667	6 x 6	3.000
2 x 12	2.000	6 x 8	4.000
3 x 3	0.750	6 x 10	5.000
3 x 4	1.000	6 x 12	6.000
3 x 6	1.500	8 x 8	5.333
3 x 8	2.000	8 x 10	6.667
3 x 10	2.500	8 x 12	8.000
3 x 12	3.000		

Factors for the Board Foot Measure of Floor or Ceiling Joists

This table allows you to compute the amount of lumber (in board feet) for the floor or ceiling joists of a known area, based on the size of lumber and the spacing. To use this chart, simply multiply the area of the room or building in question by the factor across from the appropriate board size and spacing.

Ceiling Joists			
Joist Size	Inches On Center	Board Feet per Square Foot of Area	Approximate Nails Lbs. per MBM
2″ x 4″	12″	.78	17
	16″	.59	19
	20″	.48	19
2″ x 6″	12″	1.15	11
	16″	.88	13
	20″	.72	13
	24″	.63	13
2″ x 8″	12″	1.53	9
	16″	1.17	9
	20″	.96	9
	24″	.84	9
2″ x 10″	12″	1.94	7
	16″	1.47	7
	20″	1.21	7
	24″	1.04	7
3″ x 8″	12″	2.32	6
	16″	1.76	6
	20″	1.44	6
	24″	1.25	6

EXAMPLE:
You have a 10′ x 20′ area that you have to use 2″ x 6″ floor joists. To calculate the number of board feet you will enter the 2″ x 6″ section of the table. If your spacing is to be 16″ O.C. your calculations will be:

$$10' \times 20' \times .88 \ \frac{\text{B.F.}}{\text{S.F.}} = 176 \text{ B.F.}$$

You will also need approximately

$$176 \text{ B.F.} \times \frac{13 \text{ Lbs.}}{100 \text{ B.F.}} = 2\ 1/4 \text{ Lbs. of nails}$$

Pieces of Studding Required for Walls, Floors, or Ceilings

This table lists the number of pieces of studding or furring needed for framing walls, floors, ceilings, partitions, etc. based on the length of the item being framed and its required spacing. Note that no provisions have been made for the doubling of studs at corners, window openings, door openings, etc. These must be added as called for.

Length of Wall, Floor or Ceiling (in feet)	On Center Spacing			
	12″	16″	20″	24″
8	9	7	6	5
9	10	8	6	6
10	11	9	7	6
11	12	9	8	7
12	13	10	8	7
13	14	11	9	8
14	15	12	9	8
15	16	12	10	9
16	17	13	11	9
17	18	14	11	10
18	19	15	12	10
19	20	15	12	11
20	21	16	13	11
21	22	17	14	12
22	23	18	14	12
23	24	18	15	13
24	25	19	15	13
25	26	20	16	14
26	27	21	17	14
27	28	21	17	15
28	29	22	18	15
29	30	23	18	16
30	31	24	19	16
32	33	25	20	17
34	35	27	22	18
36	37	28	23	19

EXAMPLE:
You have to frame a partition 18′ long 10′ high with 2″ x 4″ lumber at 16″ spacing.

You will need 15 pieces of 10′ long 2″ x 4″s. You should then add for blocking/bridging etc.

(excerpted from *How to Estimate Building Losses and Construction Costs*, Paul I. Thomas, Prentice-Hall)

Multipliers for Computing Required Studding or Furring

This table allows you to compute the number of pieces of studding or furring required for framing walls, floors, roofs, partitions, etc.

Spacing (in inches)	Factor by Which to Multiply the Width of Room, Length of Wall or Roof
12	1.00
16	.75
18	.67
20	.60
24	.50

Add one for end in each case.

Examples:	Answers:
A room 16′ wide requires studs 16″ O.C.	16′ x .75 + 1 = 13 joists
A wall 30′ long requires studs 18″ O.C.	30′ x .67 + 1 = 21 studs
A gable roof 40′ long requires studs 24″ O.C.	40′ x .50 + 1 = 21 rafters

(excerpted from *How to Estimate Building Losses and Construction Costs*, Paul I. Thomas, Prentice Hall)

Partition Framing

This table can be used to compute the board feet of lumber required for each square foot of wall area to be framed, based on the lumber size and spacing design.

Studs Including Sole and Cap Plates				Horizontal Bracing in All Partitions		Horizontal Bracing In Bearing Partitions Only	
Stud Size	Inches On Center	Board Feet per Square Foot of Partition Area	Lbs. Nails per MBM of Stud Framing	Board Feet per Square Foot of Partition Area	Lbs. Nails per MBM of Bracing	Board Feet per Square Foot of Partition Area	Lbs. Nails per MBM of Bracing
2″ x 3″	12″	.91	25	.04	145	.01	145
	16″	.83	25	.04	111	.01	111
	20″	.78	25	.04	90	.01	90
	24″	.76	25	.04	79	.01	79
2″ x 4″	12″	1.22	19	.05	108	.02	108
	16″	1.12	19	.05	87	.02	87
	20″	1.05	19	.05	72	.02	72
	24″	1.02	19	.05	64	.02	64
2″ x 6″	16″	1.38	19			.04	59
	20″	1.29	16			.04	48
	24″	1.22	16			.04	43
2″ x 4″ Staggered	8″	1.69	22				
3″ x 4″	16″	1.35	17				
2″ x 4″ 2″ Way	16″	1.08	19				

Exterior Wall Stud Framing

This table allows you to compute the board feet of lumber required for each square foot of exterior wall to be framed based on the lumber size and spacing design.

		Studs Including Corner Bracing		Horizontal Bracing Midway Between Plates	
Stud Size	Inches On Center	Board Feet per Square Foot of Ext. Wall Area	Lbs. of Nails per MBM of Stud Framing	Board Feet per Square Foot of Ext. Wall Area	Lbs. of Nails per MBM of Bracing
2″ x 3″	16″	.78	30	.03	117
	20″	.74	30	.03	97
	24″	.71	30	.03	85
2″ x 4″	16″	1.05	22	.04	87
	20″	.98	22	.04	72
	24″	.94	22	.04	64
2″ x 6″	16″	1.51	15	.06	59
	20″	1.44	15	.06	48
	24″	1.38	15	.06	43

Floor Framing

This table can be used to compute the board feet required for each square foot of floor area based on the size of lumber to be used and the spacing design.

Floor Joists				Block Over Main Bearing	
Joist Size	Inches On Center	Board Feet per Square Foot of Floor Area	Nails Lbs. per MBM	Board Feet per Square Foot of Floor Area	Nails Lbs. per MBM Blocking
2″ x 6″	12″	1.28	10	.16	133
	16″	1.02	10	.03	95
	20″	.88	10	.03	77
	24″	.78	10	.03	57
2″ x 8″	12″	1.71	8	.04	100
	16″	1.36	8	.04	72
	20″	1.17	8	.04	57
	24″	1.03	8	.05	43
2″ x 10″	12″	2.14	6	.05	70
	16″	1.71	6	.05	57
	20″	1.48	6	.06	46
	24″	1.30	6	.06	34
2″ x 12″	12″	2.56	5	.06	66
	16″	2.05	5	.06	47
	20″	1.77	5	.07	39
	24″	1.56	5	.07	29
3″ x 8″	12″	2.56	5	.04	39
	16″	2.05	5	.05	57
	20″	1.77	5	.06	45
	24″	1.56	5	.06	33
3″ x 10″	12″	3.20	4	.05	72
	16″	2.56	4	.07	46
	20″	2.21	4	.07	36
	24″	1.95	4	.08	26

Furring Quantities

This table provides multiplication factors for converting square feet of wall area requiring furring into board feet of lumber. These figures are based on lumber size and spacing.

Size	Board Feet per Square Feet of Wall Area — Spacing Center to Center				Lbs. Nails per MBM of Furring
	12″	16″	20″	24″	
1″ x 2″	.18	.14	.11	.10	55
1″ x 3″	.28	.21	.17	.14	37

Conversion Factors for Wood Joists and Rafters

This table is used to compute actual lengths or board feet of lumber for roofs of various inclines. To compute actual rafter quantities for incline roofs, multiply the quantities figured for a flat roof by the factors as shown in the table. Please note that this table DOES NOT INCLUDE QUANTITIES FOR CANTILEVERED OVERHANGS.

Roof Slope	Approximate Angle	Factor	Roof Slope	Approximate Angle	Factor
Flat	0°	1.000	12 in 12	45.0°	1.414
1 in 12	4.8°	1.003	13 in 12	47.3°	1.474
2 in 12	9.5°	1.014	14 in 12	49.4°	1.537
3 in 12	14.0°	1.031	15 in 12	51.3°	1.601
4 in 12	18.4°	1.054	16 in 12	53.1°	1.667
5 in 12	22.6°	1.083	17 in 12	54.8°	1.734
6 in 12	26.6°	1.118	18 in 12	56.3°	1.803
7 in 12	30.3°	1.158	19 in 12	57.7°	1.873
8 in 12	33.7°	1.202	20 in 12	59.0°	1.943
9 in 12	36.9°	1.250	21 in 12	60.3°	2.015
10 in 12	39.8°	1.302	22 in 12	61.4°	2.088
11 in 12	42.5°	1.357	23 in 12	62.4°	2.162

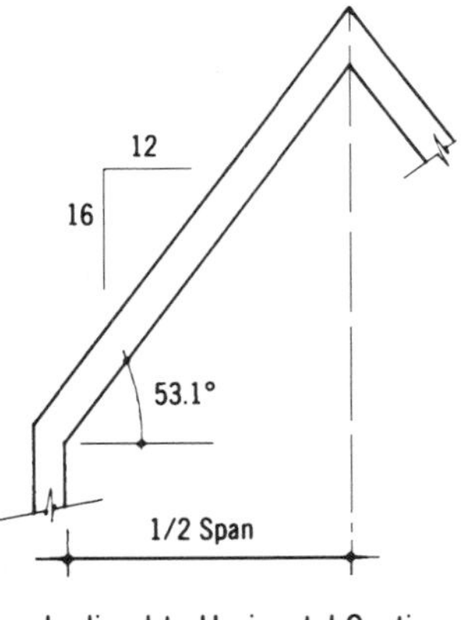

Inclined to Horizontal Sections

Roofing Factors When Rate of Rise is Known

No. Inches Rise per Foot of Run	*Pitch	Rafter Length in Inches per Foot of Run	To Obtain Rafter Length Multiply Run (in feet) By:
3.0	1/8	12.37 ÷ 12 =	1.030
3.5			1.045
4.0	1/6	12.65	1.060
4.5			1.075
5.0	5/24	13.00	1.090
5.5			1.105
6.0	1/4	13.42	1.120
6.5			1.140
7.0	7/24	13.89	1.160
7.5			1.185
8.0	1/3	14.42	1.201
8.5			1.230
9.0	3/8	15.00	1.250
9.5			1.280
10.0	5/12	16.62	1.301
10.5			1.335
11.0	11/24	16.28	1.360
11.5			1.390
12.0	1/2	16.97	1.420

*Pitch is determined by dividing the inches rise-per-foot-of-run by 24. (Most roofs of dwellings are constructed with a pitch of 1/8, 1/6, 1/4, 1/3 or 1/2.)

(excerpted from *How to Estimate Building Losses and Construction Costs*, Paul I. Thomas, Prentice Hall)

Flat Roof Framing

This table can be used to compute the amount of lumber in board feet required to frame a flat roof, based on the size lumber required.

Flat Roof Framing			
Joist Size	Inches On Center	Board Feet per Square Foot of Ceiling Area	Nails Lbs. per MBM
2″ x 6″	12″	1.17	10
	16″	.91	10
	20″	.76	10
	24″	.65	10
2″ x 8″	12″	1.56	8
	16″	1.21	8
	20″	1.01	8
	24″	.86	8
2″ x 10″	12″	1.96	6
	16″	1.51	6
	20″	1.27	6
	24″	1.08	6
2″ x 12″	12″	2.35	5
	16″	1.82	5
	20″	1.52	5
	24″	1.30	5
3″ x 8″	12″	2.35	5
	16″	1.82	5
	20″	1.52	5
	24″	1.30	5
3″ x 10″	12″	2.94	4
	16″	2.27	4
	20″	1.90	4
	24″	1.62	4

MBM; MFBM = Thousand Feet Board Measure

Allowance Factors for Roof Overhangs

Horizontal Span	Roof Overhang Measured Horizontally							
	0′-6″	1′-0″	1′-6″	2′-0″	2′-6″	3′-0″	3′-6″	4′-0″
6′	1.083	1.167	1.250	1.333	1.417	1.500	1.583	1.667
7′	1.071	1.143	1.214	1.286	1.357	1.429	1.500	1.571
8′	1.063	1.125	1.188	1.250	1.313	1.375	1.438	1.500
9′	1.056	1.111	1.167	1.222	1.278	1.333	1.389	1.444
10′	1.050	1.100	1.150	1.200	1.250	1.300	1.350	1.400
11′	1.045	1.091	1.136	1.182	1.227	1.273	1.318	1.364
12′	1.042	1.083	1.125	1.167	1.208	1.250	1.292	1.333
13′	1.038	1.077	1.115	1.154	1.192	1.231	1.269	1.308
14′	1.036	1.071	1.107	1.143	1.179	1.214	1.250	1.286
15′	1.033	1.067	1.100	1.133	1.167	1.200	1.233	1.267
16′	1.031	1.063	1.094	1.125	1.156	1.188	1.219	1.250
17′	1.029	1.059	1.088	1.118	1.147	1.176	1.206	1.235
18′	1.028	1.056	1.083	1.111	1.139	1.167	1.194	1.222
19′	1.026	1.053	1.079	1.105	1.132	1.158	1.184	1.211
20′	1.025	1.050	1.075	1.100	1.125	1.150	1.175	1.200
21′	1.024	1.048	1.071	1.095	1.119	1.143	1.167	1.190
22′	1.023	1.045	1.068	1.091	1.114	1.136	1.159	1.182
23′	1.022	1.043	1.065	1.087	1.109	1.130	1.152	1.174
24′	1.021	1.042	1.063	1.083	1.104	1.125	1.146	1.167
25′	1.020	1.040	1.060	1.080	1.100	1.120	1.140	1.160
26′	1.019	1.038	1.058	1.077	1.096	1.115	1.135	1.154
27′	1.019	1.037	1.056	1.074	1.093	1.111	1.130	1.148
28′	1.018	1.036	1.054	1.071	1.089	1.107	1.125	1.143
29′	1.017	1.034	1.052	1.069	1.086	1.103	1.121	1.138
30′	1.017	1.033	1.050	1.067	1.083	1.100	1.117	1.133
32′	1.016	1.031	1.047	1.063	1.078	1.094	1.109	1.125

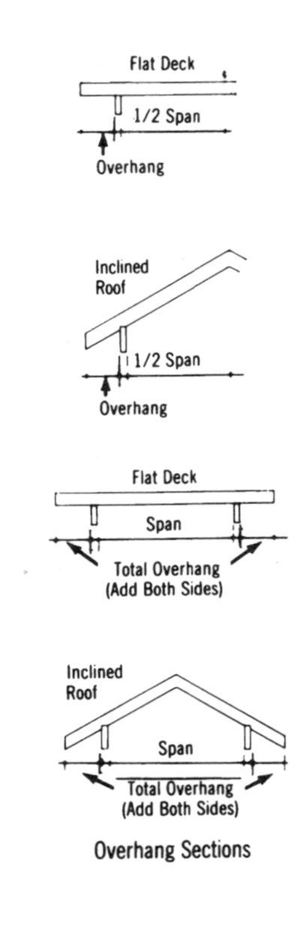

Overhang Sections

Pitched Roof Framing

This table is used to compute the board feet of lumber required per square foot of roof area, based on the required spacing and the lumber size to be used.

Rafters Including Collar Ties, Hip and Valley Rafters, Ridge Poles								
	Spacing Center to Center							
	12″		16″		20″		24″	
Rafter Size	Board Feet per Square Foot of Roof Area	Nails Lbs. per MBM	Board Feet per Square Foot of Roof Area	Nails Lbs. per MBM	Board Feet per Square Foot of Roof Area	Nails Lbs. per MBM	Board Feet per Square Foot of Roof Area	Nails Lbs. per MBM
2″ x 4″	.89	17	.71	17	.59	17	.53	17
2″ x 6″	1.29	12	1.02	12	.85	12	.75	12
2″ x 8″	1.71	9	1.34	9	1.12	9	.98	9
2″ x 10″	2.12	7	1.66	7	1.38	7	1.21	7
2″ x 12″	2.52	6	1.97	6	1.64	6	1.43	6
3″ x 8″	2.52	6	1.97	6	1.64	6	1.43	6
3″ x 10″	3.13	5	2.45	5	2.02	5	1.78	5

Material Required for On-the-Job Cut Bridging—Lineal Feet

Based on the size of the lumber used for joists, the total amount of lumber (in board feet) can be obtained for the various spacing of the joists using this table.

Size Joist in Inches	Spacing in Inches	Lineal Feet per Set (2)	Lineal Feet per Foot-of-Row
2 x 6	16	2.57	1.92
2 x 8	16	2.70	2.02
2 x 10	16	2.87	2.15
2 x 12	16	3.06	2.30
2 x 8	20	3.31	2.00
2 x 10	20	3.45	2.07
2 x 12	20	3.61	2.17
2 x 8	24	3.94	2.00
2 x 10	24	4.05	1.97
2 x 12	24	4.19	2.10

Note: Add to the total lineal feet developed from the table at least 10% cutting waste.

Example: A room 20 feet wide has two rows of bridging. The 2″ x 12″ joists are 16″ on center.

2 x 20 L.F. x 2.30 ft. per foot-of-row = 92.0 L.F.

Add 10% waste 9.2

Total L.F. = 101.2

Round out to 101 L.F.

(Per set method would be, 30 sets x 3.06 = 91.8 to which must be added 10% waste.)

(excerpted from *How to Estimate Building Losses and Construction Costs*, Paul I. Thomas, Prentice Hall)

Board Feet Required for On-the-Job Cut Bridging

Based on the size of the lumber used for joists, the total lengths of lumber for various sized bridging can be obtained from this chart.

Cross Bridging—Board Feet per Square Foot of Floors, Ceiling or Flat Roof Area Nails—Pounds Per MBM of Bridging							
		1″ x 3″		1″ x 4″		2″ x 3″	
Joist Size	Spacing	B.F.	Nails	B.F.	Nails	B.F.	Nails
2″ x 8″	12″	.04	147	.05	112	.08	77
	16″	.04	120	.05	91	.08	61
	20″	.04	102	.05	77	.08	52
	24″	.04	83	.05	63	.08	42
2″ x 10″	12″	.04	136	.05	103	.08	71
	16″	.04	114	.05	87	.08	58
	20″	.04	98	.05	74	.08	50
	24″	.04	80	.05	61	.08	41
2″ x 12″	12″	.04	127	.05	96	.08	67
	16″	.04	108	.05	82	.08	55
	20″	.04	94	.05	71	.08	48
	24″	.04	78	.05	59	.08	39
3″ x 8″	12″	.04	160	.05	122	.08	84
	16″	.04	127	.05	96	.08	66
	20″	.04	107	.05	81	.08	54
	24″	.04	86	.05	65	.08	44
3″ x 10″	12″	.04	146	.05	111	.08	77
	16″	.04	120	.05	91	.08	62
	20″	.04	102	.05	78	.08	52
	24″	.04	83	.05	63	.08	42

Board Feet of Sheathing and Subflooring

This table is used to compute the board feet of sheathing or subflooring required per square foot of roof, ceiling or flooring.

			Diagonal			
			Lbs. Nails per MBM Lumber			
			Joist, Stud or Rafter Spacing			
Type	Size	Board Feet per Square Foot of Area	12″	16″	20″	24″
Surface 4 Sides (S4S)	1″ x 4″	1.22	58	46	39	32
	1″ x 6″	1.18	39	31	25	21
	1″ x 8″	1.18	30	23	19	16
	1″ x 10″	1.17	35	27	23	19
Tongue and Groove (T&G)	1″ x 4″	1.36	65	51	43	36
	1″ x 6″	1.26	42	33	27	23
	1″ x 8″	1.22	31	24	20	17
	1″ x 10″	1.20	36	28	24	19
Shiplap	1″ x 4″	1.41	67	53	45	37
	1″ x 6″	1.29	43	33	28	23
	1″ x 8″	1.24	31	24	20	17
	1″ x 10″	1.21	36	28	24	19

Wood Siding Factors

This table shows the factor by which area to be covered is multiplied to determine exact amount of surface material needed.

Item	Nominal Size	Width Overall	Face	Area Factor
Shiplap	1″ x 6″	5-1/2″	5-1/8″	1.17
	1 x 8	7-1/4	6-7/8	1.16
	1 x 10	9-1/4	8-7/8	1.13
	1 x 12	11-1/4	10-7/8	1.10
Tongue and Groove	1 x 4	3-3/8	3-1/8	1.28
	1 x 6	5-3/8	5-1/8	1.17
	1 x 8	7-1/8	6-7/8	1.16
	1 x 10	9-1/8	8-7/8	1.13
	1 x 12	11-1/8	10-7/8	1.10
S4S	1 x 4	3-1/2	3-1/2	1.14
	1 x 6	5-1/2	5-1/2	1.09
	1 x 8	7-1/4	7-1/4	1.10
	1 x 10	9-1/4	9-1/4	1.08
	1 x 12	11-1/4	11-1/4	1.07
Solid paneling	1 x 6	5-7/16	5-7/16	1.19
	1 x 8	7-1/8	6-3/4	1.19
	1 x 10	9-1/8	8-3/4	1.14
	1 x 12	11-1/8	10-3/4	1.12
Bevel Siding*	1 x 4	3-1/2	3-1/2	1.60
	1 x 6	5-1/2	5-1/2	1.33
	1 x 8	7-1/4	7-1/4	1.28
	1 x 10	9-1/4	9-1/4	1.21
	1 x 12	11-1/4	11-1/4	1.17

Note: This area factor is strictly so-called milling waste. The cutting and fitting waste must be added.

*1″ lap

(from *Western Wood Products Association*)

Milling and Cutting Waste Factors for Wood Siding

This table lists the milling waste, the cutting waste, and the amount of nails required for various types of wood siding. Amounts are per 1,000 FBM of wood siding.

Type of Siding	Nominal Size in Inches	Lap in Inches 1" Lap	Pounds Nails per 1,000 FBM	Percentage of Waste
Bevel Siding	1 x 4	1	25-6d common	63
	1 x 6	1	25-6d common	35
	1 x 8	1-1/4	20-8d common	35
	1 x 10	1-1/2	20-8d common	30
Rustic and Drop Siding	1 x 4	Matched	40-8d common	33
	1 x 6	Matched	30-8d common	25
	1 x 8	Matched	25-8d common	20
Vertical Siding	1 x 6	Matched	25-8d finish	20
	1 x 8	Matched	20-8d finish	18
	1 x 10	Matched	20-8d finish	15
Batten Siding*	1 x 8	Rough	25	5
	1 x 10	Rough	20	5
	1 x 12	Rough	20	5
	1 x 8	Dressed	25	13
	1 x 10	Dressed	20	11
	1 x 12	Dressed	20	10
Plywood Siding	1/4	Sheets	15 per MSF	5-10
	3/8	Sheets		5-10
	5/8	Sheets		5-10

*For 1" x 10" boards, allow 1,334 lineal feet 1" x 2" joint strips for each 1,000 FBM of batten siding.
Add 12 pounds 8d common nails.
(excerpted from *How to Estimate Building Losses and Construction Costs*, Paul I. Thomas, Prentice Hall)

Nails—Dimensions

This table describes the sizes, lengths, gauge number, and approximate number of nails one can expect per pound, based on the type of nail and use.

Penny Nail System					
			Approximate Number to Pound		
Size	Length in inches	Gauge Number	Common	Finishing	Casing
2d	1	15	850		
3d	1-1/4	14	550	640	
4d	1-1/2	12-1/2	350	456	
5d	1-3/4	12-1/2	230	328	
6d	2	11-1/2	180	273	228
7d	2-1/4	11	140	170	178
8d	2-1/2	10-1/4	100	151	133
9d	2-3/4	9-1/2	80	125	100
10d	3	9	65	107	96
12d	3-1/4	9	50		60
16d	3-1/2	8	40		50
20d	4	6	31		
30d	4-1/2	5	22		
40d	5	4	18		
50d	5-1/2	3	14		
60d	6	2	12		

*For 1" x 10" boards, allow 1,334 lineal feet 1" x 2" joint strips for each 1,000 FBM of battem siding.
Add 12 pounds 8d common nails.
(excerpted from *How to Estimate Building Losses and Construction Costs*, Paul I. Thomas, Prentice Hall)

Size and Quantity of Nails for a Job

This table shows the size and approximate quantity of nails needed for various portions of a wood project.
Where a mix of nails is shown, judgment must be exercised as to the different sizes of lumber.

Kind of Framing and Size of Lumber	Size of Nails Used	Lbs. per 1,000 FBM
Sills and plates	10d, 16d & 20d	8
Wall and partition stud	10d & 16d	10
Joists and rafters		
2″ x 6″	16d, some 20d	9
2″ x 8″	16d, some 20d	8
2″ x 10″	16d, some 20d	7
2″ x 12″	16d, some 20d	6
Average for total house framing	8d, 10d, 16d, 20d	15
Wood cross bridging, 1″ x 3″	8d	1 lb. per 12 sets
Furring (100 S.F.) on masonry	8d	1
Furring (100 S.F.) on studding	8d	1/2
Roof trusses	10d, 20d and 40d	10

Rough Hardware Allowances

Average Material Cost Allowances for Rough Hardware as a Percentage of Carpentry Material Costs	
Minimum	0.5%
Maximum	1.5%

Loose Fill Insulation

Square feet covered by a 40 lb. bag of mineral wool or glass fiber. (Includes area occupied by studding or joists.) Divide the factors into the area to determine the number of bags required.

Fill Depth	Fill Density		
	6 Lbs./C.F.	8 Lbs./C.F.	10 Lbs./C.F.
1"	85.0	63.8	51.0
2"	42.5	31.9	25.5
3"	28.4	21.3	17.0
3-1/2"	24.3	18.3	14.5
3-5/8"	23.5	17.6	14.1
4"	21.2	16.0	12.7
6"	14.2	10.6	8.5
10"	8.5	6.4	5.1

(courtesy, *Estimating Tables for Home Building*, Paul I. Thomas, Craftsman Book Company)

Batt Insulation Quantities

To determine the number of insulation batts needed, multiply the factor listed for the size batt to be used by the area (square feet) to be insulated. Note: S.F. area includes that occupied by studding or joists.

Batt Sizes	No. of Batts per S.F.
15" x 24"	.38
15" x 48"	.19
19" x 24"	.30
19" x 48"	.15
23" x 24"	.25
23" x 48"	.125

(courtesy, *Estimating Tables for Home Building*, Paul I. Thomas, Craftsman Book Company)

Waste Values for Roof Styles

This table lists the approximate values for waste that should be added to the net area for materials of various roof types.

Roof Shape	Plain	Cut-up
Gable	10	15
Hip	15	20
Gambrel	10	20
Gothic	10	15
Mansard (sides)	10	15
Porches	10	–

Roofing Material Quantities — Shingles

This table describes various types of roofing shingles and some of their characteristics and expected coverages.

Product	Configuration	Approximate Shipping Weight per Square	Shingles per Square	Bundles per Square	Width	Length	Exposure	ASTM* Fire & Wind Ratings
Self-Sealing Random-Tab Strip Shingle Multi-Thickness	Various Edge, Surface Texture & Application Treatments	240# to 360#	64 to 90	3, 4 or 5	11-1/2" to 14"	36" to 40"	4" to 6"	A or C - Many Wind Resistant
Self-Sealing Random-Tab Strip Shingle Single Thickness	Various Edge, Surface Texture & Application Treatments	240# to 300#	65 to 80	3 or 4	12" to 13-1/4"	36" to 40"	4" to 5-5/8"	A or C - Many Wind Resistant
Self-Sealing Square-Tab Strip Shingle Three-Tab	3 Tab or 4 Tab	200# to 300#	65 to 80	3 or 4	12" to 13-1/4"	36" to 40"	5" to 5-5/8"	A or C - All Wind Resistant
Self-Sealing Square-Tab Strip Shingle No Cut-Out	Various Edge and Surface Texture Treatments	200# to 300#	65 to 81	3 or 4	12" to 13-1/4"	36" to 40"	5" to 5 5/8"	A or C - All Wind Resistant
Individual Interlocking Shingle Basic Design	Several Design Variations	180# to 250#	72 to 120	3 or 4	18" to 22-1/4"	20" to 22-1/2"	—	A or C - Many Wind Resistant

Other types available from some manufacturers in certain areas of the country. Consult your Regional Asphalt Roofing Manufacturers Association manufacturer.

* American Society for Testing and Materials.

(courtesy *Asphalt Roofing Manufacturers Association*)

Roofing Material Quantities — Red Cedar Shingles

This table shows expected coverages and materials required for the installation of various sizes of red cedar shingles.

Covering Capacities and Approximate Nail Requirements of Certigrade Red Cedar Shingles												
	No. 1 Grade Sixteen Inch Shingles				No. 1 Grade Eighteen Inch Shingles				No. 1 Grade Twenty-Four Inch Shingles			
	Nail Size: 3d, 1-1/4″ Long				Nail Size: 3d, 1-1/4″ Long				Nail Size: 4d, 1-1/2″ Long			
Shingles Exposure in Inches	Four-Bundle Square		One Bundle		Four Bundle: Square		One Bundle		Four-Bundle: Square		One Bundle	
	Coverage in S.F.	Pounds Nails	Coverage in S.F.	Pounds Nails	Coverage in S.F.	Pounds Nails	Coverage in S.F.	Pounds Nails	Coverage in S.F.	Pounds Nails	Coverage in S.F.	Pounds Nails
3-1/2	70	2-7/8	17-1/2	3/4								
4	80	2-1/2	20	5/8	72-1/2	2-1/2	18	5/8				
4-1/2	90	2-1/4	22-1/2	5/8	81-1/2	2-1/4	20	5/8				
5	100*	2	25	1/2	90-1/2	2	22-1/2	1/2				
5-1/2	110	1-3/4	27-1/2	1/2	100*	1-3/4	25	1/2				
6	120	1-2/3	30	3/8	109	1-2/3	27	3/8	80	2-1/3	20	5/8
6-1/2	130	1-1/2	32-1/2	3/8	118	1-1/2	29-1/2	3/8	86-1/2	2-1/8	21-1/2	1/2
7	140	1-2/5	35	1/3	127	1-2/5	31-1/2	1/3	93	2	23	1/2
7-1/2	150†	1-1/3	37-1/2	1/3	136	1-1/3	34	1/3	100*	1-7/8	25	1/2
8	160		40		145-1/2†	1-1/4	36	1/3	106-1/2	1-3/4	26-1/2	1/2
8-1/2	170		42-1/2		154-1/2	1-1/4	38-1/2	1/4	113	1-2/3	28	1/2
9	180		45		163-1/2		40-1/2		120	1-1/2	30	3/8
9-1/2	190		47-1/2		172-1/2		43		126-1/2	1-1/2	31-1/2	3/8
10	200		50		181-1/2		45		133	1-1/2	33	3/8
10-1/2	210		52-1/2		191		47-1/2		140	1-1/3	35	1/3
11	220		55		200		50		146-1/2	1-1/4	36-1/2	1/3
11-1/2	230		57-1/2		209		52		153†	1-1/4	38	1/3
12	240‡		60		218		54-1/2		160		40	
12-1/2					227		56-1/2		166-1/2		41-1/2	
13					236		59		173		43	
13-1/2					245-1/2		61		180		45	
14					254-1/2‡		63-1/2		186-1/2		46-1/2	
14-1/2									193		48	
15									200		50	
15-1/2									206-1/2		51-1/2	
16									213‡		53	

* Maximum exposure recommended for roofs.

† Maximum exposure recommended for single-coursing on side walls.

‡ Maximum exposure recommended for double-coursing on side walls. Figures in italics are inserted for convenience in estimating quantity of shingles needed for wide exposures in double-coursing, with butt-nailing. In double-coursing, with any exposure chosen, the figures indicate the amount of shingles for the outer courses. Order an equivalent number of shingles for concealed courses. Approximately 1-1/2 lbs. 5d small-headed nails required per square (100 S.F. wall area) to apply outer course of 16-inch shingles at 12-inch weather exposure. Plus 1/2 lb. 3d nails for under course shingles. Figure slightly fewer nails for 18-inch shingles at 14-inch exposure.

(from *Certigrade Handbook of Red Cedar Shingles*, published by Red Cedar Shingle Bureau (Red Cedar Shingle and Handsplit Shake Bureau), Seattle, Washington, 1957.)

(courtesy of *American Institute of Steel Construction, Inc.*)

Roofing Material Quantities — Standard 3/16″ Thick Slate

This table shows the materials needed for installing various sizes of 3/16″-thick roof slates.

Size of Slate (In.)	Slates per Square	Exposure with 3″ Lap	Nails per Square	
			Lbs.	Ozs.
26 x 14	89	11-1/2″	1	0
24 x 16	86	10-1/2″	1	0
24 x 14	98	10-1/2″	1	2
24 x 13	106	10-1/2″	1	3
24 x 12	114	10-1/2″	1	5
24 x 11	125	10-1/2″	1	7
22 x 14	108	9-1/2″	1	4
22 x 13	117	9-1/2″	1	5
22 x 12	126	9-1/2″	1	7
22 x 11	138	9-1/2″	1	9
22 x 10	152	9-1/2″	1	12
20 x 14	121	8-1/2″	1	6
20 x 13	132	8-1/2″	1	8
20 x 12	141	8-1/2″	1	10
20 x 11	154	8-1/2″	1	12
20 x 10	170	8-1/2″	1	15
20 x 9	189	8-1/2″	2	3
18 x 14	137	7-1/2″	1	9
18 x 13	148	7-1/2″	1	11
18 x 12	160	7-1/2″	1	13
18 x 11	175	7-1/2″	2	0
18 x 10	192	7-1/2″	2	3
18 x 9	213	7-1/2″	2	7

Size of Slate (In.)	Slates per Square	Exposure with 3″ Lap	Nails per Square	
			Lbs.	Ozs.
16 x 14	160	6-1/2″	1	13
16 x 12	184	6-1/2″	2	2
16 x 11	201	6-1/2″	2	5
16 x 10	222	6-1/2″	2	8
16 x 9	246	6-1/2″	2	13
16 x 8	277	6-1/2″	3	2
14 x 12	218	5-1/2″	2	8
14 x 11	238	5-1/2″	2	11
14 x 10	261	5-1/2″	3	3
14 x 9	291	5-1/2″	3	5
14 x 8	327	5-1/2″	3	12
14 x 7	374	5-1/2″	4	4
12 x 10	320	4-1/2″	3	10
12 x 9	355	4-1/2″	4	1
12 x 8	400	4-1/2″	4	9
12 x 7	457	4-1/2″	5	3
12 x 6	533	4-1/2″	6	1
11 x 8	450	4″	5	2
11 x 7	515	4″	5	14
10 x 8	515	3-1/2″	5	14
10 x 7	588	3-1/2″	7	4
10 x 6	686	3-1/2″	7	

Roofing Material Quantities — Asphalt Rolls and Sheets

This table shows some of the characteristics of asphalt roofing materials.

Product	Approximate Shipping Weight		Squares per Package	Length	Width	Selvage	Exposure	ASTM Fire & Wind Ratings
	per Roll	per Square						
Mineral Surface Roll	75# to 90#	75# to 90#	1	36′ to 38′	36″	0″ to 4″	32″ to 34″	C
Mineral Surface Roll Double Coverage	55# to 70#	110# to 140#	1/2	36′	36″	19″	17″	C
Smooth Surface Roll	50# to 86#	40# to 65#	1 to 2	36′ to 72′	36″	N/A	34″	None
Saturated Felt Underlayment (non-perforated)	35# to 60#	11# to 30#	2 to 4	72′ to 144	36″	NA	17″ to 34″	*

*May be component in a complete fire-rated system. Check with manufacturer.
(courtesy *Asphalt Roofing Manufacturers Association*)

Wood Siding Quantities

This table allows you to compute the amount of wood siding (in board feet required per square foot of wall area to receive siding) based on the style and dimensions of the siding.

Type of Siding	Size	Exposure	Board Feet per S.F. of Wall Area	Lbs. Nails per MBM of Siding — Stud Spacing 16"	20"	24"
Plain Bevel Siding	1/2" x 4"	2-1/2"	1.60	17	13	11
		2-3/4"	1.45			
	1/2" x 6"	4-1/2"	1.33	11	9	7
		4-3/4"	1.26			
		5"	1.20			
	1/2" x 8"	6-1/2"	1.23	8	7	6
		7"	1.14			
Plain Bevel Bungalow Siding	5/8" x 8"	6-1/2"	1.23	14	11	9
		7"	1.14			
	5/8" x 10"	8-1/2"	1.18	16	13	11
		9"	1.11			
	3/4" x 8"	6-1/2"	1.23	14	11	9
		7"	1.14			
	3/4" x 10"	8-1/2"	1.18	16	13	11
		9"	1.11			
	3/4" x 12"	10-1/2"	1.14	14	11	9
		11"	1.09			
Drop or Rustic Siding	3/4" x 4"	3-1/4"	1.23	27	22	18
	3/4" x 6"	5-1/6"	1.19	18	15	12
		5-3/16"	1.17			

Standard Door Nomenclature

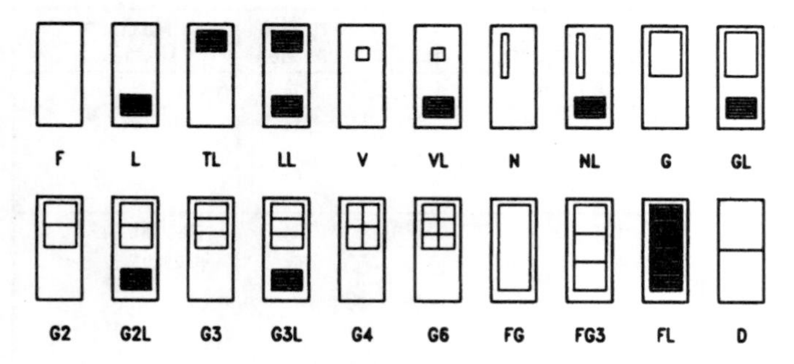

NOMENCLATURE LETTER SYMBOLS

F = Flush

L = Louvered (bottom)

TL = Louvered (top)

LL = Louvered (top and bottom)

V = Vision Lite

VL = Vision Lite and Louvered

N = Narrow Lite

NL = Narrow Lite and Louvered

G = Half Glass (options G2, G3, G4, and G6)

GL = Half Glass and Louvered (options G2L and G3L)

FG = Full Glass (option FG3)

FLI = Full Louver (Inserted)

D = Dutch Door

Weight of Doors

Door Thickness	Weight of Doors in Pounds per Square Foot				
	White Pine	Oak	Hollow Core	Solid Core	Hollow Metal
1-3/8″	3 psf	6 psf	1-1/2 psf	3-1/2 – 4 psf	6-1/2 psf
1-3/4″	3-1/2	7	2	4-1/2 – 5-1/4	6-1/2
2-1/4″	4-1/2	9	–	5-1/2 – 6-3/4	6-1/2

Finish Hardware Allowances

Average Allowances for Finish Hardware as a Percentage of Total Job Costs	
Minimum	0.75%
Maximum	3.50%

Average distribution of total finish hardware costs for a typical building is 85% material, 15% labor.

Fire Door Classifications

This table lists the various fire door ratings by label, time, and temperature rating, plus allowable glass area for a fire-rated door.

Classification	Time Rating (as shown on label)		Temperature Rise (as shown on label)	Maximum Glass Area
3 Hour fire doors (A) are for use in openings in walls separating buildings or dividing a single building into fire areas.	3 Hr.	(A)	30 Min. 250°F Max	None
	3 Hr.	(A)	30 Min. 450°F Max	
	3 Hr.	(A)	30 Min. 650°F Max	
	3 Hr.	(A)	*	
1-1/2 Hour Fire Doors (B) and (D) are for use in openings in 2 Hour enclosures of vertical communication through buildings (stairs, elevators, etc.) or in exterior walls which are subject to severe fire exposure from outside of the building. 1 Hour fire doors (B) are for use in openings in 1 Hour enclosures of vertical communication through buildings (stairs, elevators, etc.)	1-1/2 Hr.	(B)	30 Min. 250°F Max	100 square inches per door
	1-1/2 Hr.	(B)	30 Min. 450°F Max	
	1-1/2 Hr.	(B)	30 Min. 650°F Max	
	1-1/2 Hr.	(B)	*	
	1 Hr.	(B)	30 Min. 250°F Max	
	1-1/2 Hr.	(D)	30 Min. 250°F Max	None
	1-1/2 Hr.	(D)	30 Min. 450°F Max	
	1-1/2 Hr.	(D)	30 Min. 650°F Max	
	1-1/2 Hr.	(D)	*	
3/4 Hour fire doors (C) and (E) are for use in openings in corridor and room partitions or in exterior walls which are subject to moderate fire exposure from outside of the building.	3/4 Hr.	(C)	**	1296 square
	3/4 Hr.	(E)	**	720 square inches per light
1/2 Hour fire doors and 1/3 Hour fire doors are for use where smoke control is a primary consideration and are for the protection of openings in partitions between a habitable room and a corridor when the wall has a fire-resistance rating of not more than one hour.	1/2 Hr.		**	No Limit
	1/3		**	

* The labels do not record any temperature rise limits. This means that the temperature rise on the unexposed face of the door at the end of 30 minutes of test is in excess of 650°F.

** Temperature rise is not recorded

Quantities for Perlite Gypsum Plaster

This table shows the approximate number of sacks of prepared perlite gypsum plaster required to cover 100 square yards of surface area. The coverage depends on the type of lathing or base used and the required thickness of plaster.

Type of Base	Total Plaster Thickness (inches)	Number of Sacks
Wood lath	5/8	22 (80 lbs. per sack)
Metal lath	5/8	35 (80 lbs. per sack)
Gypsum lath	1/2	120 (80 lbs. per sack)
Masonry walls	5/8	24 (67 lbs. per sack)

(courtesy *How to Estimate Building Losses and Construction Costs*, Paul I. Thomas, Prentice-Hall, Inc.)

Quantities for White Coat Finish Plaster

This table shows the approximate quantities of materials required to cover 100 square yards of surface area with a 1/16" coat finish.

Type of Finish	100 Lb. Sacks Neat Gypsum	100 Lb. Sacks Keene's Cement	50 Lb. Sacks Hydrated Lime
Lime putty	2	–	8
Keene's cement	–	4	4

Based on ASA mix of 1 (100-lb.) sack neat gypsum gauging plaster to 4 (50-lb) sacks hydrated lime, and 1 (100-lb.) sack of Keene's cement to 1 (50-lb.) sack hydrated lime for medium hard finish.

(courtesy *How to Estimate Building Losses and Construction Costs*, Paul I. Thomas, Prentice-Hall, Inc.)

Quantities for Portland Cement Plaster

This table shows the approximate quantities of materials required to cover 100 square yards of surface area with Portland cement plaster. The coverage depends on the base material used and the required thickness of plaster.

Plaster Thickness (inches)	Cubic Yards on Masonry Base	Cubic Yards on Wire Lath Base
1/4	0.75	0.90
1/2	1.50	1.80
5/8	1.37	1.64
3/4	2.25	2.70
1	3.00	3.60

(courtesy *How to Estimate Building Losses and Construction Costs*, Paul I. Thomas, Prentice Hall, Inc.)

Quantities for Job-Mixed Plaster

This table presents approximate quantities required to prepare enough job-mixed sanded or perlite plaster to cover 100 square yards of surface area. The coverage depends on the type of base material used and the required thickness of plaster.

Type of Base	Total Plaster Thickness Inches	100 Lb. Sacks of Neat Gypsum	C.Y. Sand 1:2-1/2 Mix	Aggregate Perlite		
				C.F.	4 C.F. Sacks	Mix
Wood lath	5/8	11	1.1	27.5	12.5	1:2-1/2
Metal lath	5/8	20	2.0	50.0	12.5	1:2-1/2
Gypsum lath	1/2	10	1.0	25	6.5	1:2-1/2
Masonry walls	5/8	12	1.2	30	7.5	1:3

ASA permits 250 lbs. damp loose sand or 2-1/2 C.F. of vermiculite or perlite, provided this proportioning is used for both scratch and brown coats on three-coat work. The number of 4 C.F. sacks of perlite are shown to nearest 1/2 sack.

(courtesy *How to Estimate Building Losses and Construction Costs*, Paul I. Thomas, Prentice-Hall, Inc.

Furring Quantities

This table provides multiplication factors for converting square feet of wall area requiring furring into board feet of lumber. These figures are based on lumber size and spacing.

Size	Board Feet per Square Feet of Wall Area — Spacing Center to Center				Lbs. Nails per MBM of Furring
	12"	16"	20"	24"	
1" x 2"	.18	.14	.11	.10	55
1" x 3"	.28	.21	.17	.14	1' 3-3/4"

Gypsum Panel Coverage

This table shows the expected coverage from the size panels indicated across the top of the table. Inversely, if the area to be covered with gypsum board panels is known (in square feet), you can quickly determine the required number of sheets of drywall. Note that no provision has been made for openings.

No. of Panels	Sizes and Numbers of Panels and S.F. of Area Covered							
	4' x 7'	4' x 8'	4' x 9'	4' x 10'	4' x 11'	4' x 12'	4' x 13'	4' x 14'
10	280	320	360	400	440	480	520	560
11	308	352	396	440	484	528	572	616
12	336	384	432	480	528	576	624	672
13	364	416	468	520	572	624	676	728
14	392	448	504	560	616	672	728	784
15	420	480	540	600	660	720	780	840
16	448	512	576	640	704	768	832	896
17	476	544	612	680	748	816	884	952
18	504	576	648	720	792	864	936	1008
19	532	608	684	760	836	912	988	1064
20	560	640	720	800	880	960	1040	1120
21	588	672	756	840	924	1008	1092	1176
22	616	704	792	880	968	1056	1144	1232
23	644	736	828	920	1012	1104	1196	1288
24	672	768	864	960	1056	1152	1248	1344
25	700	800	900	1000	1100	1200	1300	1400
26	728	832	936	1040	1144	1248	1352	1456
27	756	864	972	1080	1188	1296	1404	1512
28	784	896	1008	1120	1232	1344	1456	1568
29	812	928	1044	1160	1276	1392	1508	1624
30	840	960	1080	1200	1320	1440	1560	1680
31	868	992	1116	1240	1364	1488	1612	1736

Drywall Accessories

This table provides the approximate quantities of nails, joint compound and tape needed for the indicated amounts of drywall.

With this Amount of Sheetrock Gypsum Panel	Type GWB-54 Nails Required* USE	this Amount of Powder-Type Compound OR	this Amount of USG Ready-to-Use Compound All Purpose AND	this Amount of Perf-A-Tape Reinforcing Tape
100 S.F.	.6 lbs.	6 lbs.	1 Gal.	37 Ft.
200 S.F.	1.1 lbs.	12 lbs.	2 Gals.	74 Ft.
300 S.F.	1.6 lbs.	18 lbs.	2 Gals.	111 Ft.
400 S.F.	2.1 lbs.	24 lbs.	3 Gals.	148 Ft.
500 S.F.	2.7 lbs.	30 lbs.	3 Gals.	185 Ft.
600 S.F.	3.2 lbs.	36 lbs.	4 Gals.	222 Ft.
700 S.F.	3.7 lbs.	42 lbs.	5 Gals.	259 Ft.
800 S.F.	4.2 lbs.	48 lbs.	5 Gals.	296 Ft.
900 S.F.	4.8 lbs.	54 lbs.	6 Gals.	333 Ft.
1000 S.F.	5.3 lbs.	60 lbs.	6 Gals.	370 Ft.

* Spaced 7″ on ceiling; 8″ on wall.

Wallboard Quantities

This table lists factors that can be applied to predetermined area figures in order to quantify actual material requirements. Included are waste factors for the intended use of the materials. The table also lists the amount of nails required for the wallboard, based on stud spacing.

Includes Fiber Board, Gypsum Board, and Plywood, Used as Underflooring, Sheathing, Plaster Base, or as Drywall Finish					
Factors		Nails			
Used for Underflooring, Sheathing, and Plaster Base	Used for Exposed Drywall Finish	Pounds of Nails per 1000 Sq. Ft. of Wallboard — Joist, Stud or Rafter Spacing			
		12″	16″	20″	24″
1.05	1.10	7	6	5	4

Location Factors

Costs shown in *Means cost data publications* are based on National Averages for materials and installation. To adjust these costs to a specific location, simply multiply the base cost by the factor for that city. The data is arranged alphabetically by state and postal zip code numbers. For a city not listed, use the factor for a nearby city with similar economic characteristics.

STATE/ZIP	CITY	Residential	Commercial
ALABAMA			
350-352	Birmingham	.86	.87
354	Tuscaloosa	.81	.79
355	Jasper	.77	.78
356	Decatur	.80	.81
357-358	Huntsville	.82	.83
359	Gadsden	.80	.81
360-361	Montgomery	.83	.81
362	Anniston	.74	.75
363	Dothan	.80	.78
364	Evergreen	.80	.78
365-366	Mobile	.82	.83
367	Selma	.80	.78
368	Phenix City	.83	.81
369	Butler	.79	.77
ALASKA			
995-996	Anchorage	1.26	1.25
997	Fairbanks	1.25	1.24
998	Juneau	1.25	1.24
999	Ketchikan	1.31	1.30
ARIZONA			
850,853	Phoenix	.92	.89
852	Mesa/Tempe	.87	.85
855	Globe	.88	.85
856-857	Tucson	.90	.87
859	Show Low	.90	.86
860	Flagstaff	.93	.89
863	Prescott	.90	.86
864	Kingman	.90	.86
865	Chambers	.89	.85
ARKANSAS			
716	Pine Bluff	.80	.80
717	Camden	.71	.71
718	Texarkana	.75	.74
719	Hot Springs	.70	.70
720-722	Little Rock	.81	.81
723	West Memphis	.80	.80
724	Jonesboro	.80	.80
725	Batesville	.76	.76
726	Harrison	.77	.77
727	Fayetteville	.69	.67
728	Russellville	.78	.75
729	Fort Smith	.82	.79
CALIFORNIA			
900-902	Los Angeles	1.08	1.08
903-905	Inglewood	1.06	1.06
906-908	Long Beach	1.07	1.07
910-912	Pasadena	1.06	1.06
913-916	Van Nuys	1.08	1.08
917-918	Alhambra	1.08	1.08
919-921	San Diego	1.10	1.06
922	Palm Springs	1.10	1.06
923-924	San Bernardino	1.09	1.05
925	Riverside	1.12	1.08
926-927	Santa Ana	1.09	1.06
928	Anaheim	1.11	1.09
930	Oxnard	1.14	1.09
931	Santa Barbara	1.10	1.07
932-933	Bakersfield	1.11	1.06
934	San Luis Obispo	1.14	1.08
935	Mojave	1.09	1.05
936-938	Fresno	1.12	1.08
939	Salinas	1.12	1.12
940-941	San Francisco	1.20	1.23
942,956-958	Sacramento	1.11	1.10
943	Palo Alto	1.14	1.17
944	San Mateo	1.14	1.17
945	Vallejo	1.12	1.15
946	Oakland	1.15	1.18
947	Berkeley	1.15	1.18
948	Richmond	1.14	1.17
949	San Rafael	1.25	1.19
950	Santa Cruz	1.15	1.14
951	San Jose	1.22	1.20
952	Stockton	1.13	1.09
953	Modesto	1.14	1.10
CALIFORNIA (CONT'D)			
954	Santa Rosa	1.14	1.17
955	Eureka	1.11	1.10
959	Marysville	1.10	1.09
960	Redding	1.10	1.09
961	Susanville	1.11	1.10
COLORADO			
800-802	Denver	.99	.95
803	Boulder	.89	.85
804	Golden	.97	.93
805	Fort Collins	.98	.92
806	Greeley	.91	.85
807	Fort Morgan	.98	.92
808-809	Colorado Springs	.94	.92
810	Pueblo	.94	.92
811	Alamosa	.90	.88
812	Salida	.90	.88
813	Durango	.88	.86
814	Montrose	.86	.84
815	Grand Junction	.91	.86
816	Glenwood Springs	.96	.91
CONNECTICUT			
060	New Britain	1.04	1.05
061	Hartford	1.04	1.05
062	Willimantic	1.04	1.05
063	New London	1.05	1.04
064	Meriden	1.03	1.04
065	New Haven	1.04	1.05
066	Bridgeport	1.02	1.05
067	Waterbury	1.06	1.06
068	Norwalk	1.01	1.05
069	Stamford	1.04	1.08
D.C.			
200-205	Washington	.93	.95
DELAWARE			
197	Newark	1.00	1.01
198	Wilmington	1.00	1.01
199	Dover	1.00	1.01
FLORIDA			
320,322	Jacksonville	.84	.83
321	Daytona Beach	.88	.87
323	Tallahassee	.76	.78
324	Panama City	.71	.72
325	Pensacola	.85	.83
326,344	Gainesville	.85	.82
327-328,347	Orlando	.88	.86
329	Melbourne	.91	.90
330-332,340	Miami	.84	.86
333	Fort Lauderdale	.84	.86
334,349	West Palm Beach	.87	.84
335-336,346	Tampa	.81	.83
337	St. Petersburg	.82	.84
338	Lakeland	.80	.82
339,341	Fort Myers	.80	.80
342	Sarasota	.80	.81
GEORGIA			
300-303,399	Atlanta	.84	.89
304	Statesboro	.65	.67
305	Gainesville	.71	.75
306	Athens	.78	.82
307	Dalton	.68	.67
308-309	Augusta	.77	.79
310-312	Macon	.81	.81
313-314	Savannah	.81	.82
315	Waycross	.74	.74
316	Valdosta	.76	.76
317	Albany	.77	.79
318-319	Columbus	.79	.79
HAWAII			
967	Hilo	1.26	1.22
968	Honolulu	1.27	1.23

Location Factors

STATE/ZIP	CITY	Residential	Commercial
STATES & POSS.			
969	Guam	1.43	1.38
IDAHO			
832	Pocatello	.94	.93
833	Twin Falls	.80	.79
834	Idaho Falls	.84	.83
835	Lewiston	1.09	1.01
836-837	Boise	.95	.94
838	Coeur d'Alene	.96	.89
ILLINOIS			
600-603	North Suburban	1.10	1.09
604	Joliet	1.07	1.06
605	South Suburban	1.09	1.08
606	Chicago	1.12	1.11
609	Kankakee	.99	.99
610-611	Rockford	1.05	1.04
612	Rock Island	1.06	.97
613	La Salle	1.05	.98
614	Galesburg	1.06	.99
615-616	Peoria	1.07	1.01
617	Bloomington	1.04	.99
618-619	Champaign	1.04	1.00
620-622	East St. Louis	.99	.99
623	Quincy	.97	.95
624	Effingham	1.00	.97
625	Decatur	1.00	.97
626-627	Springfield	1.01	.98
628	Centralia	.98	.98
629	Carbondale	.96	.96
INDIANA			
460	Anderson	.94	.92
461-462	Indianapolis	.97	.95
463-464	Gary	1.03	1.01
465-466	South Bend	.93	.91
467-468	Fort Wayne	.90	.91
469	Kokomo	.92	.91
470	Lawrenceburg	.92	.89
471	New Albany	.92	.88
472	Columbus	.95	.92
473	Muncie	.93	.92
474	Bloomington	.96	.93
475	Washington	.93	.93
476-477	Evansville	.95	.95
478	Terre Haute	.95	.94
479	Lafayette	.91	.91
IOWA			
500-503,509	Des Moines	.97	.92
504	Mason City	.87	.81
505	Fort Dodge	.84	.79
506-507	Waterloo	.88	.83
508	Creston	.90	.85
510-511	Sioux City	.95	.89
512	Sibley	.80	.78
513	Spencer	.81	.79
514	Carroll	.85	.80
515	Council Bluffs	.95	.89
516	Shenandoah	.83	.78
520	Dubuque	.99	.89
521	Decorah	.89	.80
522-524	Cedar Rapids	1.00	.92
525	Ottumwa	.95	.87
526	Burlington	.92	.87
527-528	Davenport	.98	.96
KANSAS			
660-662	Kansas City	.96	.94
664-666	Topeka	.86	.85
667	Fort Scott	.86	.84
668	Emporia	.82	.81
669	Belleville	.87	.81
670-672	Wichita	.89	.86
673	Independence	.82	.79
674	Salina	.85	.81
675	Hutchinson	.80	.76
676	Hays	.85	.81
677	Colby	.85	.81
678	Dodge City	.85	.82
679	Liberal	.78	.76
KENTUCKY			
400-402	Louisville	.95	.92
403-405	Lexington	.88	.85

STATE/ZIP	CITY	Residential	Commercial
KENTUCKY (CONT'D)			
406	Frankfort	.93	.87
407-409	Corbin	.78	.73
410	Covington	.98	.95
411-412	Ashland	.97	.98
413-414	Campton	.77	.74
415-416	Pikeville	.83	.84
417-418	Hazard	.77	.73
420	Paducah	.97	.92
421-422	Bowling Green	.96	.91
423	Owensboro	.91	.89
424	Henderson	.96	.94
425-426	Somerset	.76	.72
427	Elizabethtown	.94	.90
LOUISIANA			
700-701	New Orleans	.87	.86
703	Thibodaux	.86	.86
704	Hammond	.84	.83
705	Lafayette	.85	.82
706	Lake Charles	.83	.83
707-708	Baton Rouge	.83	.82
710-711	Shreveport	.81	.81
712	Monroe	.79	.79
713-714	Alexandria	.78	.78
MAINE			
039	Kittery	.77	.79
040-041	Portland	.88	.90
042	Lewiston	.89	.90
043	Augusta	.79	.79
044	Bangor	.92	.92
045	Bath	.77	.77
046	Machias	.83	.83
047	Houlton	.82	.82
048	Rockland	.83	.83
049	Waterville	.79	.78
MARYLAND			
206	Waldorf	.88	.88
207-208	College Park	.90	.90
209	Silver Spring	.90	.90
210-212	Baltimore	.92	.92
214	Annapolis	.89	.90
215	Cumberland	.88	.88
216	Easton	.73	.74
217	Hagerstown	.90	.88
218	Salisbury	.77	.78
219	Elkton	.83	.84
MASSACHUSETTS			
010-011	Springfield	1.04	1.02
012	Pittsfield	.98	.98
013	Greenfield	1.01	.99
014	Fitchburg	1.08	1.04
015-016	Worcester	1.10	1.06
017	Framingham	1.06	1.07
018	Lowell	1.08	1.08
019	Lawrence	1.08	1.08
020-022, 024	Boston	1.14	1.15
023	Brockton	1.06	1.08
025	Buzzards Bay	1.02	1.04
026	Hyannis	1.05	1.06
027	New Bedford	1.06	1.07
MICHIGAN			
480,483	Royal Oak	1.02	1.01
481	Ann Arbor	1.05	1.04
482	Detroit	1.07	1.06
484-485	Flint	.99	1.00
486	Saginaw	.96	.97
487	Bay City	.96	.97
488-489	Lansing	1.01	.98
490	Battle Creek	1.01	.94
491	Kalamazoo	1.00	.93
492	Jackson	1.00	.97
493,495	Grand Rapids	.89	.86
494	Muskegon	.95	.92
496	Traverse City	.88	.85
497	Gaylord	.88	.89
498-499	Iron Mountain	.98	.95
MINNESOTA			
550-551	Saint Paul	1.13	1.10
553-555	Minneapolis	1.15	1.12

Location Factors

STATE/ZIP	CITY	Residential	Commercial
556-558	Duluth	1.06	1.07
559	Rochester	1.06	1.03
560	Mankato	1.02	1.01
561	Windom	.92	.91
562	Willmar	.95	.94
563	St. Cloud	1.13	1.05
564	Brainerd	1.08	1.01
565	Detroit Lakes	.90	.97
566	Bemidji	.93	1.00
567	Thief River Falls	.89	.96
MISSISSIPPI			
386	Clarksdale	.70	.67
387	Greenville	.81	.78
388	Tupelo	.71	.72
389	Greenwood	.72	.69
390-392	Jackson	.81	.78
393	Meridian	.77	.76
394	Laurel	.73	.69
395	Biloxi	.85	.81
396	Mccomb	.69	.67
397	Columbus	.71	.72
MISSOURI			
630-631	St. Louis	1.00	1.03
633	Bowling Green	.92	.95
634	Hannibal	1.00	.94
635	Kirksville	.87	.91
636	Flat River	.95	.98
637	Cape Girardeau	.93	.96
638	Sikeston	.91	.93
639	Poplar Bluff	.90	.92
640-641	Kansas City	1.03	1.00
644-645	St. Joseph	.89	.93
646	Chillicothe	.82	.86
647	Harrisonville	.98	.96
648	Joplin	.84	.86
650-651	Jefferson City	.98	.92
652	Columbia	.99	.93
653	Sedalia	.99	.92
654-655	Rolla	.96	.90
656-658	Springfield	.85	.87
MONTANA			
590-591	Billings	.98	.96
592	Wolf Point	.98	.96
593	Miles City	.97	.95
594	Great Falls	.97	.96
595	Havre	.96	.95
596	Helena	.97	.96
597	Butte	.96	.95
598	Missoula	.95	.94
599	Kalispell	.94	.93
NEBRASKA			
680-681	Omaha	.92	.91
683-685	Lincoln	.88	.83
686	Columbus	.74	.73
687	Norfolk	.84	.83
688	Grand Island	.88	.83
689	Hastings	.82	.78
690	McCook	.80	.76
691	North Platte	.87	.82
692	Valentine	.78	.74
693	Alliance	.75	.72
NEVADA			
889-891	Las Vegas	1.06	1.05
893	Ely	.94	.95
894-895	Reno	.95	1.00
897	Carson City	.96	.99
898	Elko	.92	.94
NEW HAMPSHIRE			
030	Nashua	.93	.94
031	Manchester	.93	.94
032-033	Concord	.92	.93
034	Keene	.79	.80
035	Littleton	.82	.83
036	Charleston	.77	.78
037	Claremont	.77	.77
038	Portsmouth	.92	.91

STATE/ZIP	CITY	Residential	Commercial
NEW JERSEY			
070-071	Newark	1.14	1.12
072	Elizabeth	1.11	1.09
073	Jersey City	1.13	1.12
074-075	Paterson	1.12	1.12
076	Hackensack	1.11	1.11
077	Long Branch	1.11	1.09
078	Dover	1.12	1.10
079	Summit	1.10	1.08
080,083	Vineland	1.12	1.08
081	Camden	1.12	1.09
082,084	Atlantic City	1.11	1.08
085-086	Trenton	1.13	1.11
087	Point Pleasant	1.12	1.10
088-089	New Brunswick	1.12	1.10
NEW MEXICO			
870-872	Albuquerque	.89	.91
873	Gallup	.90	.92
874	Farmington	.90	.92
875	Santa Fe	.89	.91
877	Las Vegas	.89	.91
878	Socorro	.89	.91
879	Truth/Consequences	.88	.88
880	Las Cruces	.86	.86
881	Clovis	.91	.91
882	Roswell	.92	.92
883	Carrizozo	.92	.92
884	Tucumcari	.91	.91
NEW YORK			
100-102	New York	1.34	1.34
103	Staten Island	1.29	1.29
104	Bronx	1.28	1.28
105	Mount Vernon	1.17	1.17
106	White Plains	1.16	1.16
107	Yonkers	1.20	1.20
108	New Rochelle	1.17	1.17
109	Suffern	1.15	1.15
110	Queens	1.29	1.29
111	Long Island City	1.30	1.30
112	Brooklyn	1.30	1.30
113	Flushing	1.31	1.31
114	Jamaica	1.30	1.30
115,117,118	Hicksville	1.25	1.25
116	Far Rockaway	1.31	1.31
119	Riverhead	1.27	1.27
120-122	Albany	.97	.97
123	Schenectady	.98	.98
124	Kingston	1.11	1.09
125-126	Poughkeepsie	1.14	1.12
127	Monticello	1.09	1.07
128	Glens Falls	.95	.93
129	Plattsburgh	.95	.93
130-132	Syracuse	.98	.96
133-135	Utica	.91	.94
136	Watertown	.93	.96
137-139	Binghamton	.94	.94
140-142	Buffalo	1.06	1.02
143	Niagara Falls	1.06	1.02
144-146	Rochester	.99	1.00
147	Jamestown	.97	.94
148-149	Elmira	.96	.94
NORTH CAROLINA			
270,272-274	Greensboro	.75	.76
271	Winston-Salem	.75	.76
275-276	Raleigh	.77	.77
277	Durham	.75	.76
278	Rocky Mount	.69	.69
279	Elizabeth City	.70	.70
280	Gastonia	.74	.75
281-282	Charlotte	.74	.75
283	Fayetteville	.76	.76
284	Wilmington	.73	.75
285	Kinston	.68	.68
286	Hickory	.67	.67
287-288	Asheville	.73	.75
289	Murphy	.67	.68
NORTH DAKOTA			
580-581	Fargo	.79	.84
582	Grand Forks	.78	.83
583	Devils Lake	.77	.82
584	Jamestown	.77	.82
585	Bismarck	.81	.85

Location Factors

STATE/ZIP	CITY	Residential	Commercial
NORTH DAKOTA (CONT'D)			
586	Dickinson	.78	.82
587	Minot	.80	.84
588	Williston	.77	.81
OHIO			
430-432	Columbus	.98	.96
433	Marion	.91	.92
434-436	Toledo	1.00	.99
437-438	Zanesville	.92	.91
439	Steubenville	.98	.98
440	Lorain	1.04	.98
441	Cleveland	1.10	1.03
442-443	Akron	1.01	1.00
444-445	Youngstown	1.00	.97
446-447	Canton	.97	.96
448-449	Mansfield	.97	.95
450	Hamilton	1.00	.94
451-452	Cincinnati	1.01	.95
453-454	Dayton	.94	.93
455	Springfield	.95	.93
456	Chillicothe	1.02	.96
457	Athens	.88	.87
458	Lima	.95	.94
OKLAHOMA			
730-731	Oklahoma City	.82	.83
734	Ardmore	.83	.82
735	Lawton	.85	.84
736	Clinton	.80	.82
737	Enid	.83	.82
738	Woodward	.82	.81
739	Guymon	.69	.69
740-741	Tulsa	.85	.82
743	Miami	.86	.83
744	Muskogee	.76	.73
745	Mcalester	.76	.78
746	Ponca City	.82	.81
747	Durant	.79	.81
748	Shawnee	.79	.81
749	Poteau	.85	.81
OREGON			
970-972	Portland	1.09	1.07
973	Salem	1.07	1.06
974	Eugene	1.07	1.06
975	Medford	1.06	1.05
976	Klamath Falls	1.07	1.06
977	Bend	1.07	1.06
978	Pendleton	1.04	1.02
979	Vale	.99	.97
PENNSYLVANIA			
150-152	Pittsburgh	1.03	1.01
153	Washington	1.02	1.00
154	Uniontown	1.01	.99
155	Bedford	1.03	.97
156	Greensburg	1.02	1.00
157	Indiana	1.05	.98
158	Dubois	1.04	.97
159	Johnstown	1.04	.98
160	Butler	1.01	.98
161	New Castle	1.01	.98
162	Kittanning	1.02	.99
163	Oil City	.91	.96
164-165	Erie	.99	.97
166	Altoona	1.04	.96
167	Bradford	1.00	.98
168	State College	.97	.97
169	Wellsboro	.94	.95
170-171	Harrisburg	.97	.96
172	Chambersburg	.96	.95
173-174	York	.97	.95
175-176	Lancaster	.96	.94
177	Williamsport	.92	.91
178	Sunbury	.96	.95
179	Pottsville	.96	.95
180	Lehigh Valley	1.02	1.01
181	Allentown	1.02	1.01
182	Hazleton	.97	.96
183	Stroudsburg	1.03	1.02
184-185	Scranton	.96	.99
186-187	Wilkes-Barre	.93	.96
188	Montrose	.94	.97
189	Doylestown	.95	1.06

STATE/ZIP	CITY	Residential	Commercial
PENNSYLVANIA (CONT'D)			
190-191	Philadelphia	1.13	1.11
193	Westchester	1.07	1.05
194	Norristown	1.09	1.07
195-196	Reading	.97	.98
PUERTO RICO			
009	San Juan	.88	.88
RHODE ISLAND			
028	Newport	1.02	1.04
029	Providence	1.02	1.04
SOUTH CAROLINA			
290-292	Columbia	.72	.75
293	Spartanburg	.71	.74
294	Charleston	.74	.76
295	Florence	.71	.73
296	Greenville	.71	.74
297	Rock Hill	.65	.67
298	Aiken	.65	.68
299	Beaufort	.68	.70
SOUTH DAKOTA			
570-571	Sioux Falls	.88	.82
572	Watertown	.84	.78
573	Mitchell	.84	.78
574	Aberdeen	.85	.79
575	Pierre	.85	.79
576	Mobridge	.85	.78
577	Rapid City	.85	.79
TENNESSEE			
370-372	Nashville	.86	.86
373-374	Chattanooga	.82	.81
375,380-381	Memphis	.86	.86
376	Johnson City	.80	.79
377-379	Knoxville	.80	.80
382	Mckenzie	.69	.69
383	Jackson	.68	.75
384	Columbia	.76	.76
385	Cookeville	.68	.68
TEXAS			
750	Mckinney	.89	.82
751	Waxahackie	.82	.82
752-753	Dallas	.89	.85
754	Greenville	.78	.73
755	Texarkana	.88	.77
756	Longview	.84	.74
757	Tyler	.91	.80
758	Palestine	.73	.73
759	Lufkin	.77	.77
760-761	Fort Worth	.84	.83
762	Denton	.88	.79
763	Wichita Falls	.80	.80
764	Eastland	.74	.73
765	Temple	.78	.77
766-767	Waco	.81	.80
768	Brownwood	.73	.72
769	San Angelo	.79	.75
770-772	Houston	.87	.88
773	Huntsville	.74	.74
774	Wharton	.76	.77
775	Galveston	.86	.87
776-777	Beaumont	.82	.84
778	Bryan	.81	.82
779	Victoria	.79	.79
780	Laredo	.77	.78
781-782	San Antonio	.82	.83
783-784	Corpus Christi	.81	.80
785	Mc Allen	.79	.77
786-787	Austin	.79	.82
788	Del Rio	.68	.68
789	Giddings	.73	.72
790-791	Amarillo	.81	.81
792	Childress	.76	.78
793-794	Lubbock	.78	.80
795-796	Abilene	.77	.77
797	Midland	.78	.79
798-799,885	El Paso	.80	.79
UTAH			
840-841	Salt Lake City	.91	.90
842,844	Ogden	.91	.89

Location Factors

STATE/ZIP	CITY	Residential	Commercial
UTAH (CONT'D)			
843	Logan	.92	.90
845	Price	.82	.81
846-847	Provo	.91	.90
VERMONT			
050	White River Jct.	.73	.72
051	Bellows Falls	.74	.73
052	Bennington	.72	.71
053	Brattleboro	.74	.74
054	Burlington	.85	.86
056	Montpelier	.83	.84
057	Rutland	.86	.85
058	St. Johnsbury	.75	.75
059	Guildhall	.73	.74
VIRGINIA			
220-221	Fairfax	.89	.90
222	Arlington	.89	.90
223	Alexandria	.90	.91
224-225	Fredericksburg	.84	.85
226	Winchester	.79	.79
227	Culpeper	.79	.80
228	Harrisonburg	.75	.76
229	Charlottesville	.84	.82
230-232	Richmond	.86	.84
233-235	Norfolk	.83	.83
236	Newport News	.83	.82
237	Portsmouth	.82	.82
238	Petersburg	.86	.84
239	Farmville	.75	.73
240-241	Roanoke	.77	.76
242	Bristol	.80	.75
243	Pulaski	.73	.72
244	Staunton	.77	.75
245	Lynchburg	.81	.77
246	Grundy	.73	.73
WASHINGTON			
980-981,987	Seattle	.99	1.04
982	Everett	.97	1.03
983-984	Tacoma	1.05	1.03
985	Olympia	1.05	1.03
986	Vancouver	1.10	1.04
988	Wenatchee	.94	.98
989	Yakima	1.02	1.00
990-992	Spokane	.99	.98
993	Richland	1.01	1.00
994	Clarkston	1.00	.99
WEST VIRGINIA			
247-248	Bluefield	.90	.90
249	Lewisburg	.91	.91
250-253	Charleston	.94	.94
254	Martinsburg	.76	.77
255-257	Huntington	.95	.97
258-259	Beckley	.92	.92
260	Wheeling	.93	.95
261	Parkersburg	.94	.96
262	Buckhannon	.97	.94
263-264	Clarksburg	.97	.94
265	Morgantown	.97	.94
266	Gassaway	.94	.94
267	Romney	.92	.92
268	Petersburg	.96	.93
WISCONSIN			
530,532	Milwaukee	1.01	1.00
531	Kenosha	1.01	1.00
534	Racine	1.05	1.00
535	Beloit	1.00	.98
537	Madison	1.00	.98
538	Lancaster	.92	.90
539	Portage	.99	.96
540	New Richmond	1.03	.95
541-543	Green Bay	1.00	.97
544	Wausau	.99	.95
545	Rhinelander	.99	.95
546	La Crosse	.98	.95
547	Eau Claire	1.04	.96
548	Superior	1.04	.98
549	Oshkosh	.97	.94

STATE/ZIP	CITY	Residential	Commercial
WYOMING			
820	Cheyenne	.86	.82
821	Yellowstone Nat. Pk.	.81	.78
822	Wheatland	.83	.79
823	Rawlins	.82	.78
824	Worland	.79	.76
825	Riverton	.82	.79
826	Casper	.87	.83
827	Newcastle	.80	.76
828	Sheridan	.84	.81
829-831	Rock Springs	.83	.79
CANADIAN FACTORS (reflect Canadian currency)			
ALBERTA			
	Calgary	1.00	.97
	Edmonton	1.00	.97
	Fort McMurray	.99	.96
	Lethbridge	.99	.96
	Lloydminster	.99	.96
	Medicine Hat	.99	.96
	Red Deer	.99	.96
BRITISH COLUMBIA			
	Kamloops	1.03	1.04
	Prince George	1.05	1.06
	Vancouver	1.06	1.07
	Victoria	1.05	1.06
MANITOBA			
	Brandon	.98	.97
	Portage la Prairie	.98	.97
	Winnipeg	.97	.97
NEW BRUNSWICK			
	Bathurst	.93	.91
	Dalhousie	.93	.91
	Fredericton	.96	.94
	Moncton	.93	.91
	Newcastle	.93	.91
	Saint John	.97	.95
NEWFOUNDLAND			
	Corner Brook	.95	.94
	St. John's	.95	.94
NORTHWEST TERRITORIES			
	Yellowknife	.92	.91
NOVA SCOTIA			
	Dartmouth	.97	.96
	Halifax	.97	.96
	New Glasgow	.97	.96
	Sydney	.95	.94
	Yarmouth	.97	.96
ONTARIO			
	Barrie	1.09	1.07
	Brantford	1.11	1.09
	Cornwall	1.09	1.07
	Hamilton	1.12	1.08
	Kingston	1.09	1.07
	Kitchener	1.05	1.03
	London	1.08	1.06
	North Bay	1.07	1.05
	Oshawa	1.09	1.07
	Ottawa	1.09	1.07
	Owen Sound	1.08	1.06
	Peterborough	1.08	1.06
	Sarnia	1.11	1.09
	St. Catharines	1.04	1.02
	Sudbury	1.04	1.02
	Thunder Bay	1.05	1.03
	Toronto	1.12	1.11
	Windsor	1.06	1.04
PRINCE EDWARD ISLAND			
	Charlottetown	.93	.91
	Summerside	.93	.91

Location Factors

STATE/ZIP	CITY	Residential	Commercial
QUEBEC			
	Cap-de-la-Madeleine	1.04	1.03
	Charlesbourg	1.04	1.03
	Chicoutimi	1.03	1.02
	Gatineau	1.02	1.01
	Laval	1.04	1.02
	Montreal	1.09	1.03
	Quebec	1.11	1.04
	Sherbrooke	1.04	1.02
	Trois Rivieres	1.04	1.03
SASKATCHEWAN			
	Moose Jaw	.92	.92
	Prince Albert	.92	.92
	Regina	.92	.92
	Saskatoon	.92	.92
YUKON			
	Whitehorse	.92	.91

Abbreviations

A	Area
ASTM	American Society for Testing and Materials
B.F.	Board feet
Carp.	Carpenter
C.F.	Cubic feet
CWJ	Composite wood joist
C.Y.	Cubic yard
Ea.	Each
Equip.	Equipment
Exp.	Exposure
Ext.	Exterior
F	Fahrenheit
Ft.	Foot, feet
Gal.	Gallon
Hr.	Hour
in.	Inch, inches
Inst.	Installation
Int.	Interior
Lb.	Pound
L.F.	Linear feet
LVL	Laminated veneer lumber
Mat.	Material
Max.	Maximum
MBF	Thousand board feet
MBM	Thousand feet board measure
MSF	Thousand square feet
Min.	Minimum
O.C.	On center
O&P	Overhead and profit
OWJ	Open web wood joist
Oz.	Ounce
Pr.	Pair
Quan.	Quantity
S.F.	Square foot
Sq.	Square, 100 square feet
S.Y.	Square yard
V.L.F.	Vertical linear feet
′	Foot, feet
″	Inch, inches
°	Degrees

Index

CMD Group...

R.S. Means Company, Inc., a CMD Group company, which is a division of Cahners Business Information, the leading provider of construction cost data in North America, supplies comprehensive construction cost guides, related technical publications and educational services.

CMD Group, a leading worldwide provider of total construction information solutions, is comprised of three synergistic product groups crafted to be the complete resource for reliable, timely and actionable project, product and cost data. In North America, CMD Group encompasses:

- Architects' First Source
- Construction Market Data (CMD)
- Manufacturer's Survey Associates (MSA)
- R.S. Means
- CMD Canada
- BIMSA/Mexico
- Worldwide, CMD Group includes Byggfakta Scandinavia (Denmark, Estonia, Finland, Norway and Sweden) and Cordell Building Information Services (Australia).

First Source for Products, available in print, on the Means CostWorks CD, and on the Internet, is a comprehensive product information source. In alliance with The Construction Specifications Institute (CSI) and Thomas Register, Architects' First Source also produces CSI's SPEC-DATA® and MANU-SPEC® as well as CADBlocks.℠ Together, these products offer commercial building product information for the building team at each stage of the construction process.

CMD Exchange, a single electronic community where all members of the construction industry can communicate, collaborate, and conduct business, better, faster, easier.

Construction Market Data provides complete, accurate and timely project information through all stages of construction. Construction Market Data supplies industry data through productive leads, project reports, contact lists, market penetration analysis and sales evaluation reports. Any of these products can pinpoint a county, look at a state, or cover the country. Data is delivered via paper, e-mail or the Internet.

CMD Group Canada serves the Canadian construction market with reliable and comprehensive information services that cover all facets of construction. Core services include: Buildcore, product selection and specification tools available in print and on the Internet; CMD Building Reports, a national construction project lead service; CanaData, statistical and forecasting information; Daily Commercial News, a construction newspaper reporting on news and projects in Ontario; and Journal of Commerce, reporting news in British Columbia and Alberta.

Manufacturers' Survey Associates is a quantity survey and specification service whose experienced estimating staff examines project documents throughout the bidding process and distributes edited information to its clients. Material estimates, edited plans and specifications and distribution of addenda form the heart of the Manufacturers' Survey Associates product line, which is available in print and on CD-ROM.

Clark Reports is the premier provider of industrial construction project data, noted for providing earlier, more complete project information in the planning and pre-planning stages for all segments of industrial and institutional construction.

Byggfakta Scandinavia AB, founded in 1936, is the parent company for the leaders of customized construction market data for Denmark, Estonia, Finland, Norway and Sweden. Each company fully covers the local construction market and provides information across several platforms including subscription, ad-hoc basis, electronically and on paper.

Cordell Building Information Services, with its complete range of project and cost and estimating services, is Australia's specialist in the construction information industry. Cordell provides in-depth and historical information on all aspects of construction projects and estimation, including several customized reports, construction and sales leads, and detailed cost information among others.

For more information, please visit our web site at www.cmdg.com.

CMD Group Corporate Offices
30 Technology Parkway South #100
Norcross, GA 30092-2912
(800) 793-0304
(770) 417-4002 (fax)
info@cmdg.com
www.cmdg.com

Contractor's Pricing Guides

For more information
visit Means Web Site
at http://www.rsmeans.com

Means ADA Compliance Pricing Guide

Accurately plan and budget for the ADA modifications you are most likely to need... with the first available cost guide for business owners, facility managers, and all who are involved in building modifications to comply with the Americans With Disabilities Act.

75 major projects—the most frequently needed modifications—include complete estimates with itemized materials and labor, plus contractor's total fees. Over 260 project variations fit almost any site conditions or budget constraints. Location Factors to adjust costs for 927 cities and towns.

A collaboration between Adaptive Environments Center, Inc. and R.S. Means Engineering Staff.

$72.95 per copy
Over 350 pages, illustrated, softcover
Catalog No. 67310 ISBN 0-87629-351-8

Contractor's Pricing Guide: Residential Square Foot Costs 2001

Now available in one concise volume, all you need to know to plan and budget the cost of new homes. If you are looking for a quick reference, the model home section contains costs for over 250 different sizes and types of residences, with hundreds of easily applied modifications. If you need even more detail, the Assemblies Section lets you build your own costs or modify the model costs further. Hundreds of graphics are provided, along with forms and procedures to help you get it right.

$39.95 per copy
Over 250 pages, illustrated, 8-1/2 x 11
Catalog No. 60321 ISBN 0-87629-600-2

Contractor's Pricing Guide: Residential Detailed Costs 2001

Every aspect of residential construction, from overhead costs to residential lighting and wiring, is in here. All the detail you need to accurately estimate the costs of your work with or without markups—labor-hours, typical crews and equipment are included as well. When you need a detailed estimate, this publication has all the costs to help you come up with a complete, on the money, price you can rely on to win profitable work.

$36.95 per copy
Over 300 pages, with charts and tables, 8-1/2 x 11
Catalog No. 60331 ISBN 0-87629-599-5

Contractor's Pricing Guide: Framing & Rough Carpentry 2001

The book contains prices for all aspects of framing and rough carpentry for residential and light commercial construction. Based on information provided by suppliers and wholesalers from around the country. To estimate a job, you simply match the specs to our system, calculate the price extensions, add up the total and adjust to your location with the easy-to-use location factors. It's as simple as that. All the forms you will need to prepare your estimate and present it professionally are included as well. You will also find graphics, charts, tables and checklists to help you work through the estimating process quickly and accurately.

$36.95 per copy
Over 200 pages, illustrated, 8-1/2 x 11
Catalog No. 60311 ISBN 0-87629-601-0

Means Repair & Remodeling Estimating Methods

3rd Edition

By Edward B. Wetherill and R.S. Means Engineering Staff

This updated edition focuses on the unique problems of estimating renovations in existing structures—using the latest cost resources and construction methods. The book helps you determine the true costs of remodeling, and includes:

Part I—The Estimating Process
Part II—Estimating by CSI Division
Part III—Two Complete Sample Estimates—Unit Price & Assemblies
New Section on Disaster Reconstruction

$69.95 per copy
Over 450 pages, illustrated, hardcover
Catalog No. 67265A ISBN 0-87629-454-9

Means Landscape Estimating Methods **New 3rd Edition**

By Sylvia H. Fee

Professional Methods for Estimating and Bidding Landscaping Projects and Grounds Maintenance Contracts

- Easy-to-understand text. Clearly explains the estimating process and how to use *Means Site Work & Landscape Cost Data.*
- Sample forms and worksheets to save you time and avoid errors.
- Tips on best techniques for saving money and winning jobs.
- **Two new chapters** help you control your equipment costs and bid landscape maintenance projects.

$62.95 per copy
Over 300 pages, illustrated, hardcover
Catalog No. 67295A ISBN 0-87629-534-0

Annual Cost Guides

For more information visit Means Web Site at http://www.rsmeans.com

Means Building Construction Cost Data 2001

Offers you unchallenged unit price reliability in an easy-to-use arrangement. Whether used for complete, finished estimates or for periodic checks, it supplies more cost facts better and faster than any comparable source. Over 23,000 unit prices for 2001. The City Cost Indexes now cover over 930 areas, for indexing to any project location in North America.

$89.95 per copy
Over 700 pages, softcover
Catalog No. 60011 ISBN 0-87629-581-2

Means Open Shop Building Construction Cost Data 2001

The open-shop version of the *Means Building Construction Cost Data*. More than 22,000 reliable unit cost entries based on open shop trade labor rates. Eliminates time-consuming searches for these prices. The first book with open shop labor rates and crews. Labor information is itemized by labor-hours, crew, hourly/daily output, equipment, overhead and profit. For contractors, owners and facility managers.

$89.95 per copy
Over 650 pages, softcover
Catalog No. 60151 ISBN 0-87629-590-1

Means Plumbing Cost Data 2001

Comprehensive unit prices and assemblies for plumbing, irrigation systems, commercial and residential fire protection, point-of-use water heaters, and the latest approved materials. This publication and its companion, *Means Mechanical Cost Data*, provide full-range cost estimating coverage for all the mechanical trades.

$89.95 per copy
Over 600 pages, softcover
Catalog No. 60211 ISBN 0-87629-587-1

Means Residential Cost Data 2001

Speeds you through residential construction pricing with more than 100 illustrated complete house square-foot costs. Alternate assemblies cost selections are located on adjoining pages, so that you can develop tailor-made estimates in minutes. Complete data for detailed unit cost estimates is also provided.

$79.95 per copy
Over 550 pages, softcover
Catalog No. 60171 ISBN 0-87629-597-9

Means Electrical Cost Data 2001

Pricing information for every part of electrical cost planning: unit and systems costs with design tables; engineering guides and illustrated estimating procedures; complete labor-hour, materials, and equipment costs for better scheduling and procurement. With the latest products and construction methods used in electrical work. More than 15,000 unit and systems costs, clear specifications and drawings.

$89.95 per copy
Over 460 pages, softcover
Catalog No. 60031 ISBN 0-87629-584-7

Means Repair & Remodeling Cost Data 2001

Commercial/Residential

You can use this valuable tool to estimate commercial and residential renovation and remodeling. By using the specialized costs in this manual, you'll find it's not necessary to force fit prices for new construction into remodeling cost planning. Provides comprehensive unit costs, building systems costs, extensive labor data and estimating assistance for every kind of building improvement.

$79.95 per copy
Over 600 pages, softcover
Catalog No. 60041 ISBN 0-87629-585-5

Means Site Work & Landscape Cost Data 2001

Hard-to-find costs are presented in an easy-to-use format for every type of site work and landscape construction. Costs are organized, described, and laid out for earthwork, utilities, roads and bridges, as well as grading, planting, lawns, trees, irrigation systems, and site improvements.

$89.95 per copy
Over 600 pages, softcover
Catalog No. 60281 ISBN 0-87629-588-X

Means Light Commercial Cost Data 2001

Specifically addresses the light commercial market, which is an increasingly specialized niche in the industry. Aids you, the owner/designer/contractor, in preparing all types of estimates, from budgets to detailed bids. Includes new advances in methods and materials. Assemblies section allows you to evaluate alternatives in early stages of design/planning.

$79.95 per copy
Over 600 pages, softcover
Catalog No. 60181 ISBN 0-87629-603-7

Books for Builders

For more information visit Means Web Site at http://www.rsmeans.com

Basics for Builders: Plan Reading & Material Takeoff

A complete course in reading and interpreting building plans—and performing quantity takeoffs to professional standards.

This book shows and explains, in clear language and with over 160 illustrations, typical working drawings encountered by contractors in residential and light commercial construction. The author describes not only how all common features are represented, but how to translate that information into a material list. Organized by CSI division, each chapter uses plans, details and tables, and a summary checklist.

$35.95 per copy
Over 420 pages, illustrated, softcover
Catalog No. 67307 ISBN 0-87629-348-8

Basics for Builders: How To Survive and Prosper in Construction

An easy-to-apply checklist system that helps contractors organize, manage, and market a construction firm successfully. Topics include:

- Identifying profitable markets
- Assessing the firm's capabilities
- Controlled growth
- The bid/no bid decision
- Scheduling
- Job start-up and planning
- Subcontractor management
- Finances and claims

$34.95 per copy
Over 300 pages, illustrated, softcover
Catalog No. 67273 ISBN 0-87629-342-9

Means Estimating Handbook

This comprehensive reference is for use in the field and the office. It covers a full spectrum of technical data required for estimating, with information on sizing, productivity, equipment requirements, codes, design standards and engineering factors. It will help you evaluate architectural plans and specifications; prepare accurate quantity takeoffs; perform value engineering; compare design alternatives; prepare estimates from conceptual to detail; evaluate change orders.

$99.95 per copy
Over 900 pages, hardcover
Catalog No. 67276 ISBN 0-87629-177-9

Basics for Builders: Framing & Rough Carpentry

A complete, illustrated do-it-yourself course on framing and rough carpentry. The book covers walls, floors, stairs, windows, doors, and roofs, as well as nailing patterns and procedures. Additional sections are devoted to equipment and material handling, standards, codes, and safety requirements.

The "framer-friendly" approach includes easy-to-follow, step-by-step instructions. This practical guide will benefit both the carpenter's apprentice and the experienced carpenter, and sets a uniform standard for framing crews.

$24.95 per copy
Over 125 pages, illustrated, softcover
Catalog No. 67298 ISBN 0-87629-251-1

Interior Home Improvement Costs, New 7th Edition

Estimates for 66 interior projects, including:

- Attic/Basement Conversions
- Kitchen/Bath Remodeling
- Stairs, Doors, Walls/Ceilings
- Fireplaces
- Home Offices/In-law Apartments

$19.95 per copy
Over 230 pages, illustrated, softcover
Catalog No. 67308C ISBN 0-87629-576-6

Exterior Home Improvement Costs, New 7th Edition

Quick estimates for 64 projects, including:

- Room Additions/Garages
- Roofing/Siding/Painting
- Windows/Doors
- Landscaping/Patios
- Porches/Decks

$19.95 per copy
Over 250 pages, illustrated, softcover
Catalog No. 67309C ISBN 0-87629-575-8

Practical Pricing Guides for Homeowners and Contractors

These updated resources on the cost and complexity of the nation's most popular home improvement projects include estimates of materials quantities, total project costs, and labor hours. With costs localized to over 900 zip code locations.

Books for Builders

For more information
visit Means Web Site
at http://www.rsmeans.com

Means Illustrated Construction Dictionary (Condensed Edition)

Based on *Means Illustrated Construction Dictionary, New Unabridged Edition*, the condensed version features 9,000 construction terms. If your work overlaps the construction business—from insurance, banking and real estate to building inspectors, attorneys, owners, and students—you will surely appreciate this valuable reference source.

$59.95 per copy
Over 500 pages, softcover
Catalog No. 67282 ISBN 0-87629-219-8

Superintending for Contractors:

How to Bring Jobs in On-time, On-budget

by Paul J. Cook

Today's superintendent has become a field project manager, directing and coordinating a large number of subcontractors, and overseeing the administration of contracts, change orders, and purchase orders. This book examines the complex role of the superintendent/field project manager, and provides guidelines for the efficient organization of this job.

$35.95 per copy
Over 220 pages, illustrated, softcover
Catalog No. 67233 ISBN 0-87629-272-4

Means Forms for Contractors

The most-needed forms for contractors of various-size firms and specialties.

Includes a variety of forms for each project phase -- from bidding to punch list. With sample project correspondence. Includes forms for project administration, safety and inspection, scheduling, estimating, change orders, and personnel evaluation. Blank forms are printed on heavy stock for easy photocopying. Each has a filled-in sample, with instructions and circumstances for use. 30 years of experience in construction project management, providing contractors with the tools they need to develop competitive bids. Cook's methods apply to all job sizes, up to multimillion dollar projects.

$79.95 per copy
Over 400 pages, three-ring binder
Catalog No. 67288

Estimating for Contractors:

How to Make Estimates that Win Jobs

by Paul J. Cook

This widely used reference offers clear, step-by-step estimating instructions that lead to achieving the following goals: objectivity, thoroughness, and accuracy.

Estimating for Contractors is a reference that will be used over and over, whether to check a specific estimating procedure, or to take a complete course in estimating.

$35.95 per copy
Over 225 pages, illustrated, softcover
Catalog No. 67160 ISBN 0-87629-271-6

Business Management for Contractors:

How to Make Profits in Today's Market

By Paul J. Cook

Focuses on the manager's role in ensuring that the company fulfills contracts, realizes a profit, and shows steady growth. Offers guidance on planning company growth, financial controls, and industry relations.

New reduced price
Now $17.98 per copy; limited quantity
Over 230 pages, softcover
Catalog No. 67250 ISBN 0-87629-269-4

Building Spec Homes Profitably

by Kenneth V. Johnson

The author offers a system to reduce risk and ensure profits in spec home building no matter what the economic climate. Includes:

- The 3 Keys to Success: location, floor plan and value
- Market Research: How to perform an effective analysis
- Site Selection: How to find and purchase the best properties
- Financing: How to select and arrange the best method
- Design Development: Combining value with market appeal
- Scheduling & Supervision: Expert guidance for improving your operation

$29.95 per copy
Over 200 pages, softcover
Catalog No. 67312 ISBN 0-87629-357-7

New Titles

Historic Preservation: Project Planning & Estimating

This new book combines Means' expertise in construction cost estimating with project planning and management guidance from some of the best-known authorities in historic preservation. Covers the unique requirements of historic building projects, including items such as special access and projection needs, limitations imposed by owners or funding agencies, and scheduling and coordination issues.

A major section explains how to identify and repair more than 75 historic building materials.

With guidance on what to look for in the site visits, how to create detailed estimates, and how to select qualified craftspeople.

$99.95 per copy
Over 700 pages, hardcover
Catalog No. 67323

Means Spanish-English Construction Dictionary

by R.S. Means, ICBO (International Conference Building Officials)

A valuable tool to facilitate communications among the Spanish and English-speaking communities in construction. This quick-reference can save time, increase safety, and improve quality—with definitions of the most common construction items and procedures.

In addition to 1500 terms and definitions, the dictionary includes more than 50 pages of building systems, equipment and tools—organized by trade.

Includes useful on-the-job phrases; numbers and measurements; and quantity calculations.

$22.95 per copy
Over 250 pages, softcover
Catalog No. 67327

Residential & Light Commercial Construction Standards

Compiled from Building Code Requirements, Industry Standards, and Recognized Trade Customs

Use this One-of-a-kind Reference to:

- Review recommended installation methods
- Set a standard for subcontractors and employees
- Answer client questions with an authoritative reference
- Resolve disputes and avoid litigation
- Protect against defect claims

$59.95 per copy
Over 575 pages, hundreds of illustrations
Catalog No. 67322

ORDER TOLL FREE 1-800-334-3509
OR FAX 1-800-632-6732.

2001 Order Form

Qty.	Book No.	COST ESTIMATING BOOKS	Unit Price	Total
	60061	Assemblies Cost Data 2001	$149.95	
	60011	Building Construction Cost Data 2001	89.95	
	61011	Building Const. Cost Data-Looseleaf Ed. 2001	116.95	
	63011	Building Const. Cost Data-Metric Version 2001	89.95	
	60221	Building Const. Cost Data-Western Ed. 2001	89.95	
	60111	Concrete & Masonry Cost Data 2001	83.95	
	50141	Construction Cost Indexes 2001	198.00	
	60141A	Construction Cost Index-January 2001	49.50	
	60141B	Construction Cost Index-April 2001	49.50	
	60141C	Construction Cost Index-July 2001	49.50	
	60141D	Construction Cost Index-October 2001	49.50	
	60311	Contr. Pricing Guide: Framing/Carpentry 2001	36.95	
	60331	Contr. Pricing Guide: Resid. Detailed 2001	36.95	
	60321	Contr. Pricing Guide: Resid. Sq. Ft. 2001	39.95	
	64021	ECHOS Assemblies Cost Book 2001	149.95	
	64011	ECHOS Unit Cost Book 2001	99.95	
	54001	ECHOS (Combo set of both books)	214.95	
	60231	Electrical Change Order Cost Data 2001	89.95	
	60031	Electrical Cost Data 2001	89.95	
	60201	Facilities Construction Cost Data 2001	219.95	
	60301	Facilities Maintenance & Repair Cost Data 2001	199.95	
	60161	Heavy Construction Cost Data 2001	89.95	
	63161	Heavy Const. Cost Data-Metric Version 2001	89.95	
	60091	Interior Cost Data 2001	89.95	
	60121	Labor Rates for the Const. Industry 2001	199.95	
	60181	Light Commercial Cost Data 2001	79.95	
	60021	Mechanical Cost Data 2001	89.95	
	60151	Open Shop Building Const. Cost Data 2001	89.95	
	60211	Plumbing Cost Data 2001	89.95	
	60041	Repair and Remodeling Cost Data 2001	79.95	
	60171	Residential Cost Data 2001	79.95	
	60281	Site Work & Landscape Cost Data 2001	89.95	
	60051	Square Foot Costs 2001	99.95	
		REFERENCE BOOKS		
	67147A	ADA in Practice	72.95	
	67310	ADA Pricing Guide	72.95	
	67298	Basics for Builders: Framing & Rough Carpentry	24.95	
	67273	Basics for Builders: How to Survive and Prosper	34.95	
	67307	Basics for Builders: Plan Reading & Takeoff	35.95	
	67261A	Bldg. Prof. Guide to Contract Documents-3rd Ed.	64.95	
	67312	Building Spec Homes Profitably	29.95	
	67250	Business Management for Contractors	17.98	
	67146	Concrete Repair & Maintenance Illustrated	69.95	
	67278	Construction Delays	59.95	
	67255	Contractor's Business Handbook	21.48	
	67314	Cost Planning & Est. for Facil. Maint.	82.95	
	67317A	Cyberplaces: The Internet Guide-2nd Ed.	59.95	
	67230A	Electrical Estimating Methods-2nd Ed.	64.95	
	64777	Environmental Remediation Est. Methods	99.95	
	67160	Estimating for Contractors	35.95	
	67276	Estimating Handbook	99.95	
	67249	Facilities Maintenance Management	86.95	

Qty.	Book No.	REFERENCE BOOKS (Cont.)	Unit Price	Total
	67246	Facilities Maintenance Standards	$ 79.95	
	67264	Facilities Manager's Reference	86.95	
	67318	Facilities Operations & Engineering Reference	99.95	
	67301	Facilities Planning & Relocation	89.95	
	67231	Forms for Building Const. Professional	47.48	
	67288	Forms for Contractors	79.95	
	67260	Fundamentals of the Construction Process	34.98	
	67258	Hazardous Material & Hazardous Waste	44.98	
	67148	Heavy Construction Handbook	74.95	
	67323	Historic Preservation: Proj. Planning & Est.	99.95	
	67308C	Home Improvement Costs-Interior Projects	19.95	
	67309C	Home Improvement Costs-Exterior Projects	19.95	
	67304	How to Estimate with Metric Units	19.98	
	67306	HVAC: Design Criteria, Options, Select.-2nd Ed.	84.95	
	67281	HVAC Systems Evaluation	84.95	
	67282	Illustrated Construction Dictionary, Condensed	59.95	
	67292A	Illustrated Construction Dictionary, w/CD-ROM	99.95	
	67295A	Landscape Estimating-3rd Ed.	62.95	
	67266	Legal Reference for Design & Construction	54.98	
	67299	Maintenance Management Audit	32.48	
	67302	Managing Construction Purchasing	24.98	
	67294	Mechanical Estimating-2nd Ed.	64.95	
	67245A	Planning and Managing Interior Projects-2nd Ed.	69.95	
	67283A	Plumbing Estimating Methods-2nd Ed.	59.95	
	67326	Preventive Maint. Guidelines for School Facil.	149.95	
	67236A	Productivity Standards for Constr.-3rd Ed.	99.95	
	67247A	Project Scheduling & Management for Constr.	64.95	
	67262	Quantity Takeoff for Contractors	17.98	
	67265A	Repair & Remodeling Estimating-3rd Ed.	69.95	
	67322	Residential & Light Commercial Const. Stds.	59.95	
	67254	Risk Management for Building Professionals	19.98	
	67291	Scheduling Manual-3rd Ed.	32.48	
	67145A	Square Foot Estimating Methods-2nd Ed.	69.95	
	67287	Successful Estimating Methods	64.95	
	67313	Successful Interior Projects	24.98	
	67233	Superintending for Contractors	35.95	
	67321	Total Productive Facilities Management	79.95	
	67284	Understanding Building Automation Systems	29.98	
	67259	Understanding Legal Aspects of Design/Build	79.95	
	67303	Unit Price Estimating Methods-2nd Ed.	59.95	
	67319	Value Engineering: Practical Applications	79.95	
		MA residents add 5% state sales tax		
		Shipping & Handling**		
		Total (U.S. Funds)*		

Prices are subject to change and are for U.S. delivery only. *Canadian customers may call for current prices. **Shipping & handling charges: Add 7% of total order for check and credit card payments. Add 9% of total order for invoiced orders.

Send Order To: **ADDV-1001**

Name (Please Print) ____________________

Company ____________________

☐ **Company**
☐ **Home** Address ____________________

____

City/State/Zip ____________________

Phone # ____________ P.O. # ____________

Mail To: **R.S. Means Company, Inc.,** P.O. Box 800, Kingston, MA 02364-0800 **(Must accompany all orders being billed)**